KV-337-717

UNDERDEVELOPMENT IN SPANISH AMERICA

Underdevelopment in Spanish America

AN INTERPRETATION

by

KEITH GRIFFIN
Fellow of Magdalen College Oxford

London
GEORGE ALLEN AND UNWIN LTD
RUSKIN HOUSE · MUSEUM STREET

FIRST PUBLISHED IN 1969
SECOND IMPRESSION 1971

ISBN 0 04 330150 9

PRINTED IN GREAT BRITAIN
in 10 on 11pt Times Roman
BY REDWOOD PRESS LIMITED
TROWBRIDGE & LONDON

FOR MY FATHER

whose loyalty and support in a
difficult period is gratefully remembered

'Nuestro desarrollo económico de los últimos años presenta síntesis que evidencian una situación realmente patológica. Hasta mediados del siglo XIX, el comercio exterior de Chile estaba casi exclusivamente en manos Chilenos. En menos de cincuenta años, el comercio exterior ha asfixiado nuestra incipiente iniciativa comercial; y en nuestro propio suelo nos eliminó del comercio internacional y nos desalojó, en gran parte, del comercio al detalle. . . . La marina mercante . . . ha caído en tristes dificultades y sigue cediendo campo a la navegación extranjera aún en el comercio de cabotaje. La majoría de las compañías de seguros que operan entre nosotros tienen su casa matriz en el exterior. Los bancos nacionales han cedido y siguen cediendo terreno a las sucursales de los bancos extranjeros. Una porción cada vez major de bonos de las instituciones de ahorro está pasando a manos de extranjeros que viven en el exterior.'

F. Encina, *Nuestra Inferioridad Economica*, 1912

PREFACE

In the course of studying, writing about and working in the underdeveloped countries I have become dissatisfied with most of the theories of economic development that are available and the policy recommendations that are based upon them. In particular, I have come to believe that underdevelopment is a process that is sustained by existing national and international institutions. If rapid development is to occur, these institutions must be altered.

The purpose of writing this book is to present this hypothesis in the context of one, fairly homogeneous, region—the nine Spanish speaking countries of South America. In other words, my intention is to analyse the causes of underdevelopment in Spanish America and indicate possible measures which might be adopted in order to accelerate the pace of social and economic progress. Particular attention is devoted to international economic relations and the way these influence the behaviour of the rest of the economy.

An inevitable consequence of trying to generalize about nine separate countries is that the argument has had to be conducted at a fairly high level of abstraction. While I remain convinced of the value of comparative studies and the need to synthesize from the experience of more than one case, it must be recognized that this method also has its limitations. At many points in the discussion I have been tempted to insert footnotes indicating that in one country or another conditions differ somewhat from those suggested in the text. In order to maintain a coherent approach I have tried to resist this temptation as much as possible, but the fact remains that the nine countries are not alike in all respects—even at the level of generalization which I have adopted.

It was decided at an early stage of the work that it would not be profitable to attempt to generalize about the whole of Latin America. Cuba and the other islands of the West Indies were obviously a separate case. Mexico had already experienced a major institutional transformation, although it can be argued that another one is now needed. Central America is very small—the population of all six countries is less than that of Colombia—and has its own scheme for regional integration. Brazil, on the other hand, is as large as the nine Spanish American countries combined and, although a member of the Latin American Free Trade Association, clearly does not depend on it for future industrial growth.

Some of the analysis, however, may be applicable in quite a few underdeveloped countries. The discussion in the Introduction is intended to be completely general and, indeed, I hope it will be read by those who may have no specific interest in Spanish America. The analysis of agrarian problems certainly is relevant to Central America and Brazil. Parts of the structural theory of inflation may be useful outside Latin America. The discussion of the problems of regional integration may be of interest to those concerned with African and Asian affairs as well. The issues associated with aid and trade policies do not vary very much from one region to another. In summary, I am primarily concerned with presenting a series of hypotheses and illustrating them with examples from Spanish America; I believe these hypotheses will eventually be regarded as broadly correct in the majority of underdeveloped countries.

It is a great pleasure to acknowledge the assistance I have received while preparing this volume. Any merit that may be found in this book is due in large part to Lord Balogh and Professor Paul Streeten, who between them have read and commented upon almost everything I have written in the last few years. Mr E. F. Jackson kindly examined a draft of some of the early chapters and his pertinent criticisms are appreciated. Thanks are also due to A. G. Frank, Solon Barraclough, James Becket, Lucio Geller and Victor Tokman for helpful discussions on various points. The growth model used in Chapter V was borrowed from a former teacher at Williams College, John Power. I hope he is not alarmed by the use to which his model has been put.

Much of the research that underlies this book was done at the Instituto de Economia of the University of Chile. I am grateful to its Director, Sr Carlos Massad, and his successor, Sr Roberto Maldonado, for the facilities they placed at my disposal and the encouragement they gave throughout my two years' residence. Srta Teresa Jeanneret shared her office with me and listened patiently while I discussed each new notion with her. Ricardo Ffrench-Davis has collaborated with me on numerous occasions. Many of the ideas in this volume have come under his careful scrutiny and his comments have been unusually valuable. None of the people mentioned above, however, nor any of the institutions for which I have worked, can be held responsible for my errors and opinions.

Some of the material included in this volume was previously published in *Oxford Economic Papers*, the *Bulletin* of the Oxford

University Institute of Economics and Statistics, *Inter-American Economic Affairs*, the *Journal of Common Market Studies* and the *Revista de Economia*.

K.B.G.

Magdalen College, Oxford
November 1968

CONTENTS

TABLES

Introduction

Underdevelopment in Theory and History

Most of the theorizing on economic development has been done by economists who live and were trained in the industrial West. Some economists, in fact, have written about the underdeveloped countries before they have seen them,[1] and others—although they may have visited an underdeveloped country—write as if they have seen only the capital and perhaps a few of the other major cities. Almost all of these economists, moreover, are ignorant of much of the economic history of the countries about which they are theorizing. Thus many writers on the poverty of nations have suffered from two serious handicaps: lack of knowledge about the broad historical forces associated with underdevelopment and ignorance of the institutions, behaviour responses and ways of life of the largest sector within the underdeveloped countries, the rural areas. Research now available or in progress is gradually reducing our ignorance of the causes of underdevelopment and the conditions under which most of mankind lives. It is almost certain that once additional evidence is accumulated many of the theories of development proposed in the last two decades will have to be abandoned.

THEORIES OF DUALISM

Perhaps the most pervasive theory is that of the dual economy. There are numerous models of economic dualism, but their common feature is the division of the economy into two broad—largely independent—sectors.[2] The names given to these two sectors vary.

[1] The most candid confession is by C. P. Kindleberger: 'This book is written by one who has not been there.' (*Economic Development*, McGraw-Hill, 1958, 1st edition, p. ix.)

[2] The most frequently cited authors in this literature are W. Arthur Lewis, 'Economic Development with Unlimited Supplies of Labour', *Manchester School,* May 1954, and 'Unlimited Labour: Further Notes', *Manchester School*, January

In some cases the division is between a 'capitalist' and a 'non-capitalist' sector (Lewis); in other cases it is a division between an 'enclave' and the 'hinterland', between a 'modern' and a 'traditional' sector or society[3] or, more generally, between 'industry' and 'agriculture' (Jorgenson).

The two sectors are separate and radically different. The 'modern', 'capitalist', 'industrial' sector is receptive to change, is market oriented and follows profit maximizing behaviour. The 'traditional', 'feudal', 'agricultural' sector is stagnant; production is for subsistence; little output passes through a market; the leisure preferences of producers are high and they do not follow maximizing behaviour. Unemployment, although 'disguised', is assumed to be widespread throughout the agricultural sector and, indeed, the marginal product of labour is zero if not negative.[4] Income is at a subsistence level, which is variously interpreted as either a physiological[5] or a culturally determined minimum.[6]

The methods of production are very different in the two sectors. 'The output of the traditional sector is a function of land and labour alone; there is no accumulation of capital . . .'.[7] In the manufacturing sector 'output is a function of capital and labour alone'.[8] The only link between the two sectors is a flow of unemployed labour (of homogeneous quality) from agriculture to industry. No flows of capital or savings are permitted—since production in the agricultural sector is done without the use of capital, and entrepreneurs are not allowed to engage in activities in both sectors—since motivations and behaviour in the two sectors differ. The economy is essentially

1958; J. C. H. Fei and G. Ranis, *Development of the Labour Surplus Economy: Theory and Policy*, Yale University, 1964; D. W. Jorgenson, 'The Development of a Dual Economy', *Economic Journal*, June 1961 and 'Surplus Agricultural Labour and the Development of a Dual Economy', *Oxford Economic Papers*, November 1967.

[3] See W. W. Rostow, *The Stages of Economic Growth*, Cambridge University Press, 1960.

[4] See R. Nurkse, *Problems of Capital Formation in Underdeveloped Countries*, Oxford University Press, 1957; P. N. Rosenstein-Rodan, 'Problems of Industrialization of Eastern and South Eastern Europe', *Economic Journal*, 1943; K. Mandelbaum, *The Industrialization of Backward Areas* Basil Blackwell, 1945.

[5] Leibenstein defines the subsistence level as one where 'equality between high fertility and high mortality rates exist. These are the maximum rates consistent with the survival of the population'. (H. Leibenstein, *Economic Backwardness and Economic Growth*, John Wiley and Sons, 1960, p. 154.)

[6] W. Arthur Lewis, *op. cit.*

[7] D. W. Jorgenson, 'Surplus Agricultural Labour and the Development of a Dual Economy', *loc. cit.*, p. 291.

[8] *Ibid.*, p. 292.

closed and growth occurs through a transfer of labour from agriculture to industry in response to demand generated by capitalist businessmen reinvesting their profits. This process continues until all the disguised unemployed are eliminated, labour becomes scarce and the traditional sector is forced to modernize.

Dualistic models of growth, sometimes explicitly but more often implicitly, have constituted the basis on which broad development strategies have been created. The general neglect of agriculture and the bias in favour of industry, which until recently have been such a notable feature of development policy, stem directly from these models. Moreover, within agriculture, the concentration on large commercial farmers (who may be considered to belong to the modern sector) reflects the opinion that small peasants will not respond to ordinary economic incentives. Similarly, within industry, the concentration on manufactured consumer goods which use imported inputs and the failure to take advantage of opportunities to process locally available raw materials reflects the belief that the 'traditional' sector is incapable of supplying the 'modern' sector with the inputs it requires.

The assumptions on which dualistic models are constructed are highly suspect. First, there is very little evidence of widespread unemployment throughout the year. There may indeed be pronounced seasonal unemployment in some countries,[9] although even this has been denied in at least one densely populated underdeveloped country.[10] The usual pattern, at least in countries where imports of cheap manufactured goods have not destroyed the handicraft industries, is for seasonally available rural manpower to be fully engaged in non-agricultural activities—leather work, food processing, textile spinning and weaving, etc.[11] There is little surplus labour. Secondly, the assumption that rural incomes or wages exceed the marginal product of labour (even if the latter is not zero) could be correct only if there are no commercial farming activities whatever (e.g. share cropping, fixed rental farming), no employment opportunities outside the (extended) family farm and if all farm

[9] See, for example, K. B. Griffin, 'Algerian Agriculture in Transition', *Bulletin* of the Oxford University Institute of Economics and Statistics, 1965.

[10] M. Paglin, ' "Surplus" Agricultural Labour and Development', *American Economic Review*, September 1965. Paglin's study is of India.

[11] See, for instance, M. Herskovits, *Economic Anthropology: A Study in Comparative Economics*, Knopf, 2nd edition, 1952; G. Dalton, ed., *Tribal and Peasant Economies: Readings in Economic Anthropology*, Natural History Press, 1967; J. Ingram, *Economic Change in Thailand Since 1850*, Stanford University Press, 1955.

labour is provided by members of the family.[12] Our knowledge of tenure conditions (e.g. the interdependence of latifundia and minifundia in Latin America), the role of migrant labour (e.g. in Africa) and the practice of small farmers of hiring labour during the peak of the harvest season (e.g. in Asia) contradicts the assumption of the dual economy model. In other words, there is no reason to suppose that rural labour receives more than its marginal product. If there is a discrepancy between opportunity costs and incomes this is more likely to be due to the presence of monopolistic market power than to non-maximizing behaviour or a work-and-income-sharing ethic.

The presumption that members of the 'traditional' sector of a dual economy are not maximizers is used to explain the alleged fact that labour supply curves are backward sloping and that peasants will not increase output when profit opportunities arise. In other words, dualistic models tend to suppose that if wages or farm prices increase the response will be to reduce the supply of labour or agricultural output. This view is clearly presented by a Dutch economist, J. H. Boeke:

'When the price of coconut is high, the chances are that less of the commodities will be offered for sale; when wages are raised the manager of the estate risks that less work will be done; . . . when rubber prices fall the owner of a grove may decide to tap more intensively, whereas high prices may mean that he leaves a larger or smaller portion of his tappable trees untapped.'[13]

The extraordinary thing is that there is absolutely no empirical evidence to support the view that labour will work less if paid more. Indeed there is much evidence to the contrary. It is probable that the backward bending supply curve is a myth left over from the colonial era when the colonized peoples frequently were forced to offer their services to Europeans in order to earn sufficient cash income to pay their taxes. Obviously, in such a situation, if wages are raised, taxes can be paid more easily and the volume of labour services offered to the colonialists will correspondingly decline. Thus this third assumption of dualistic models—the perverse response of workers to wage incentives—must be dismissed.

Fourth, a large number of detailed econometric studies have

[12] A useful discussion of some of these issues is found in A. Berry and R. Soligo, 'Rural-Urban Migration, Agricultural Output, and the Supply Price of Labour in a Labour-Surplus Economy', *Oxford Economic Papers*, July 1968.

[13] J. H. Boeke, *Economics and Economic Policy of Dual Societies*, New York, 1953, p 40. The backward bending supply curve of effort is also supported by B. Higgins, *Economic Development*, Constable, 1959, pp. 286–7, 504.

demonstrated beyond a doubt that the assumption that farmers in underdeveloped countries do not respond to price signals is untenable.[14] We now know that Punjabi peasants, whether Hindus in India or Muslims in Pakistan, respond actively to agricultural policies.[15] Thai farmers, under appropriate conditions, will introduce new crops and new technologies.[16] African farmers can be induced by pricing policies to improve the quality of the output of their tree crops.[17] If in many countries the rate of growth of agriculture is too low the explanation should be sought not in the motives, values and behaviour of the inhabitants of rural areas, but in land tenure conditions, in the distribution of economic power and in government policy.

Finally, the assumption that peasants cannot save because they are too poor must be questioned, even if sufficient information is not yet available to reject it completely. Nurkse presumes that savers are found 'mostly among the urban commercial classes'[18] and that 'peasants are not likely to save . . . voluntarily since they live so close to subsistence level'.[19] Lewis argues that only capitalists save and the reason savings are low in the underdeveloped countries is because the capitalist sector (and hence the proportion of income received in the form of profits) is small.[20]

Unfortunately, data to test these hypotheses are very scarce. One study of rural and urban incomes and savings habits in East and West Pakistan is worth mentioning, however, despite the fact that the quality of the statistical information is rather poor.

The two wings of Pakistan are separated by a thousand miles of

[14] See, for example, W. P. Falcon, 'Farmer Response to Price in a Subsistence Economy: The Case of West Pakistan', *American Economic Review*, May 1964; P. T. Bauer and B. S. Yamey, 'A Case Study of Response to Price in an Underdeveloped Country', *Economic Journal*, December 1959; J. R. Behrman, *Supply Response in Underdeveloped Agriculture: A Case Study of Thialand, 1937–1963;* D. Narain, *The Impact of Price Movements on Areas Under Selected Crops in India, 1900–1939*; E. Dean, *The Supply Responses of African Farmers: Theory and Measurement in Malawi*, Amsterdam, 1966.

[15] W. P. Falcon and C. H. Gotsch, 'Two Approaches with the Same Results', *Asian Review*, July 1968.

[16] J. R. Behrman, 'The Adoption of New Products and of New Factors in Response to Market Incentives in Peasant Agriculture: An Econometric Investigation of Thai Corn and Kenaf Supply Responses in the Post-war Period', University of Pennsylvania, Department of Economics, Discussion Paper No. 45, mimeo, February 1967.

[17] P. T. Bauer and B. S. Yamey, *op. cit.*

[18] R. Nurkse, *op. cit.*, p. 37.

[19] *Ibid.*, p. 43.

[20] 'Economic Development with Unlimited Supplies of Labour', *loc. cit.* Jorgenson assumes that 'saving is equal to total profits in the industrial sector' (*op. cit.*).

Indian territory. East Pakistan is the poorer of the two wings, average personal income *per capita* in 1963/64 being Rs.305 in rural areas, Rs.509 in urban areas, and Rs.316 for the Province as a whole. In West Pakistan *per capita* income was Rs.373 in rural areas, Rs.515 in urban areas and Rs.406 on average. Thus personal incomes in West Pakistan were about 28 per cent higher than in East Pakistan. What is noteworthy is that personal savings (expressed as a percentage of gross personal income before taxes) in the rural areas of Pakistan were higher than in the urban areas and that rural savings in East Pakistan were higher than in West Pakistan. When private corporate saving is added to personal saving, so as to obtain a measure of private saving, it turns out that urban areas save more than rural but that East Pakistan still saves more than West Pakistan. In general, 'rural areas . . . appear to have contributed at least three-fourths of the total private savings in the country'.[21]

Table I

PERSONAL AND PRIVATE SAVINGS IN PAKISTAN

	Gross Personal Savings as per cent of gross personal income	Gross Private Savings as per cent of gross private income
East Pakistan, rural	12·0	12·0
East Pakistan, urban	9·9	13·9
East Pakistan, combined	11·8	12·2
West Pakistan, rural	9·2	9·2
West Pakistan, urban	6·7	12·5
West Pakistan, combined	8·8	10·5
All Pakistan, rural	10·9	10·9
All Pakistan, urban	7·4	12·8
All Pakistan, combined	10·2	11·3

Source: A. Bergan, *op. cit.*, pp. 185 and 186.

Clearly one cannot reject the hypothesis that the 'traditional' sector does not save on the basis of a single study of one country in one year, but enough information has been provided to create a certain amount of doubt as to the validity of theories which are dependent upon this assumption. Indeed it now seems most unlikely that the assumptions of the model of economic dualism—and particularly the assumptions about the extent of rural unemployment, the rela-

[21] A. Bergan, 'Personal Income Distribution and Personal Savings in Pakistan, 1963/64', *Pakistan Development Review*, Summer 1967, p .186.

tionship of wages to the marginal product of labour, the willingness of peasants to save, and the response of workers and farmers to economic incentives—can withstand empirical scrutiny.

One can always maintain that the assumptions of a theory are less important than its predictions and that it is more important to foresee the development path of an economy than to describe accurately its structure and behaviour patterns. Let us, therefore, briefly examine the trends and tendencies the dual economy models would lead one to expect.

The most obvious feature of these models is the tendency for real income in the agricultural sector to remain constant. It cannot rise because there is surplus labour and it cannot fall because incomes already are at a subsistence level. Given that the marginal product of labour is zero, it must follow that all available amounts of land are fully utilized, otherwise it is very difficult to understand why the surplus labour does not combine with uncultivated land. If labour is redundant and land is fully utilized any increase in population will lead to falling *per capita* incomes—unless the increase in population is exactly offset by technical progress. The technical progress may not be of the embodied type, however, because capital accumulation in the 'traditional' sector is assumed not to occur; thus the increase in technical knowledge must be entirely disembodied, i.e. it must fall like manna from heaven at precisely the rate of population increase.

These, in fact, are the assumptions many development theorists make. Jorgenson, for example, assumes that his production function 'will shift over time so that a given bundle of factors will generate a higher level of output at one date than at an earlier date'.[22] He also assumes 'that so long as there is disguised unemployment population expands at the same rate as the growth of agricultural output'.[23] Thus *per capita* income in the agricultural sector cannot fall by assumption, and since the modern, industrial sector is assumed to be increasing its relative importance in the economy the *per capita* income of the nation as a whole must rise.

Dualistic theories thus make three specific predictions about the development path of an underdeveloped country: first, aggregate *per capita* income will rise; second, agricultural output will increase at the same rate as the population; and, third, *per capita* income in rural areas will remain constant. What evidence is there that these predictions are generally correct?

[22] 'Surplus Agricultural Labour and the Development of a Dual Economy', *loc. cit.*, p. 292.

[23] *Ibid.*, p. 293.

In the first place, there are several areas in which *per capita* income has declined. In Africa north of the Sahara, for example, gross domestic product *per capita* declined by 0·3 per cent a year between 1960 and 1967. Looking at individual countries over the period 1960–1966, GNP *per capita* grew at an annual rate of −0·1 per cent in Ghana, −0·5 per cent in Morocco, −2·6 per cent in Rhodesia, −0·4 per cent in the Dominican Republic and −1·4 per cent in Uruguay.[24] Evidently, a few countries are engaged in a process of underdevelopment which dual economy models are incapable of explaining. It is quite likely, as we shall see below, that if the economic history of today's poverty-stricken nations were examined it would become apparent that many of them descended into underdevelopment from a level of material prosperity and social wellbeing that was once considerably greater than that observed at present.

Next, there is abundant evidence that in many countries agricultural output, and particularly production of food for domestic consumption, has failed to keep pace with the rate of population increase. Comparing 1966 with the average of the period 1957–1959 it appears that *per capita* agricultural production had declined in the following countries: Algeria, Burundi, Congo (Kinshasa), Liberia, Malagasy Republic, Morocco, Rwanda, Tunisia, Uganda, Iran, Iraq, Egypt, India, Burma, Cambodia, Indonesia, South Vietnam, Argentina, Bolivia, Brazil, Chile, Colombia, Costa Rica, Dominican Republic, Ecuador, Haiti, Paraguay, Peru, Trinidad and Tobago.[25] Clearly the theories of dualism have failed this simple test on a massive scale.

Finally, there is the question of whether *per capita* incomes in rural areas have remained constant. We argue in Chapter I below that there is some evidence that rural incomes have been falling in several Spanish American countries. It is possible, of course, that our data on Spanish America are wrong or that for some reason this area constitutes a special case. Hence it is worth considering briefly what is happening in another country in a rather different part of the world.

In 1965 it was suggested, on the basis of the then existing statistical information on the *per capita* availability of foodgrains and average rural incomes, that despite the rapid growth in GNP and *per capita* income that Pakistan has enjoyed, particularly during the Second Five Year Plan, 'the vast majority of the Pakistani population

[24] AID, Statistics and Reports Division, *Economic Growth Trends: Latin-America, East Asia, Near East and South Asia, Africa*, 1967/1968.

[25] *Ibid.* Also see FAO, *The State of Food and Agriculture*, 1967.

probably have a lower standard of living today than when the country achieved its independence in 1947'.[26] Two years later additional evidence became available which showed that *per capita* agricultural output in East Pakistan declined from Rs.197 in 1949–50, to Rs.184 in 1954–55, and finally to Rs.174 in 1959–60; it then increased to Rs.188 in 1964–65 but was still lower than the level achieved in the earliest recorded period. Exactly the same pattern was followed in West Pakistan except that by 1964–65 *per capita* agricultural output had fully regained the previous peak.[27] Clearly a strong presumption exists that the standard of living in rural areas, particularly in East Pakistan, has declined in the last two decades.

The author of a more recent study has investigated in great detail the trends in rural incomes in East Pakistan since 1949.[28] According to the evidence collected by Mr Bose agricultural value added per head of rural population declined from Rs.200·5 in 1949/50–1950/51 to Rs.182 in 1962/63–1963/64; in the same period agricultural value added per head of agricultural population declined from Rs.229 to Rs.202·5 and *per capita* rural income declined from Rs.272·5 to Rs.268·5.[29] Because of increased population density in rural areas the number of landless male workers seeking wage employment rose from 14·1 per cent of the male agricultural labour force in 1951 to 19·4 per cent in 1961.[30] The real wages of these landless agricultural labourers declined from an index of 100 in 1949 to 82·3 in 1966.[31]

In view of this evidence it seems rather pointless to construct a model of an underdeveloped economy in which the central feature is the constancy of real incomes in the largest sector. Ironically, it was during the period of falling rural incomes in Pakistan that two visiting economists published the first version of their model of a dual economy based on the assumption of a fixed 'institutional or subsistence level of real wages in the agricultural sector'.[32] The Fei and Ranis model subsequently became famous, particularly in the

[26] K. B. Griffin, 'Financing Development Plans in Pakistan', *Pakistan Development Review*, Winter 1965, p. 606.

[27] G. F. Papanek, *Pakistan's Development*, Harvard University Press, 1967, Appendix Table 5B, p. 318.

[28] S. R. Bose, 'Trend of Real Incomes of the Rural Poor in East Pakistan, 1949–66—An Indirect Estimate', Pakistan Institute of Development Economics, Research Report No. 68, July 1968.

[29] *Ibid.*, Table 1. The data are expressed in constant prices of 1959/60.

[30] *Ibid.*, Table 3.

[31] *Ibid.*, Table 4.

[32] J. C. H. Fei and G. Ranis, 'Unlimited Supply of Labour and the Concept of Balanced Growth', *Pakistan Development Review*, Winter 1961, p. 32.

United States,[33] but in its country of origin it remains pathetically irrelevant.

It is conceivable, however, that some theorists would claim that models of economic dualism are not really concerned with the agricultural sector but are concerned with describing the pattern of growth of the modern industrial sector. In other words, it might be claimed that the dual economy models put a spotlight on one—often very small[34]—sector and leave the rest of the economy in relative obscurity.[35] If this is the correct interpretation, the validity of the theory should be tested by comparing its predictions with the performance of the 'modern' sector, and the 'traditional' sector should be ignored.

A prominent characteristic of the dual economy cum surplus labour theory is the lack of employment opportunities in agriculture and the growth of employment opportunities in the 'modern' sector. This is a fundamental asymmetry in the model, and it is this difference in the treatment of the growth and employment potential of the two sectors which determines the development path of the economy. Given the structure of the model it is obvious that development can occur only through a process of capital accumulation in the 'modern' sector and the absorption of labour in industry. In other words, the proportion of the labour force occupied in industry should increase, and in the classical version of the model 'the rate of growth of manufacturing employment is, of course, equal to the rate of growth of manufacturing output'.[36]

These hypotheses are contradicted by a great deal of empirical evidence. Nowhere, I believe, has industrial output and employment increased at the same rate; there has always been some increase in the productivity of labour in industry. More important, there are many countries in which the proportion of the labour force employed in industry has increased much less than the theory would lead one to

[33] See G. Ranis and J. C. H. Fei, 'A Theory of Economic Development', *American Economic Review*, September 1961; J. C. H. Fei and G. Ranis, *Development of the Labour Surplus Economy: Theory and Policy*, Yale University, 1964.

[34] In Chad the non-agricultural labour force is 8 per cent of the total, in Kenya 12 per cent, in Thailand 18 per cent, in Nepal 8 per cent, in Yemen 11 per cent and in Haiti 17 per cent.

[35] This rationalization certainly does not apply to Fei and Ranis who criticize Lewis for treating the agricultural sector 'more or less as an afterthought' while claiming that their model 'gives a more explicit treatment to the agricultural sector'. ('Unlimited Supply of Labour and the Concept of Balanced Growth', *loc. cit.*, p. 29)

[36] D. W. Jorgenson, 'Surplus Agricultural Labour and the Development of a Dual Economy', *loc. cit.*, p. 297.

expect (e.g. Turkey and Malaysia), others in which the relative size of the industrial labour force has remained constant (e.g. Egypt), and still others in which the proportion has even declined (e.g. Cyprus). As we shall see in Chapter I (Table I:3), in a few Spanish American nations there was a smaller proportion of the labour force employed in manufacturing in 1960 than in 1925.

Table 2

PERCENTAGE OF THE LABOUR FORCE EMPLOYED IN INDUSTRY, SELECTED COUNTRIES

	Circa 1950	*Circa 1960*
Nicaragua	15	16
Turkey	9	10
Pakistan	8	10
Malaysia	11	13
Egypt	12	12
Panama	12	12
Paraguay	19	19
Chile	31	30
Ecuador	23	18
Peru	21	20
Cyprus	33	27

Source: S. Baum, 'The World's Labour Force and its Industrial Distribution, 1950 and 1960', *International Labour Review*, January–February 1967, Appendix II, pp. 110–12

In a few countries, especially in Africa, there has been a decline not only in the *proportion* of the work force employed in industry and the 'modern' sector, but in the *absolute level* of employment in non-agricultural activities as well. Output of the 'modern' sector has increased while employment has fallen. For instance, between 1955 and 1964 the trend rate of increase of non-agricultural employment was −1·0 in the Cameroons, −0·5 in Kenya, −0·7 in Malawi, −0·4 in Tanzania, −0·1 in Uganda and −0·9 in Zambia.[37] The dual economy model is incapable of explaining these trends.

The theories we are examining make predictions not only about the level of employment in industry, but also about the aggregate rate of savings and investment in the economy. Arthur Lewis asserts that an understanding of how the savings and investment ratios rise from 4–5 per cent to 12–15 per cent of national income is the 'central

[37] C. R. Frank, Jr, 'Urban Unemployment and Economic Growth in Africa', *Oxford Economic Papers*, July 1968, Table II, p. 254.

problem in the theory of economic development'.[38] His explanation of the rise is in terms of a redistribution of income from the 'subsistence' to the 'capitalist' sector. Similarly, Professor Jorgenson's model implies that 'if the proportion of manufacturing output to agricultural output increases, the share of saving in total income also increases'.[39]

Once again, there is statistical evidence that suggests that there is no simple association between a growing modern, capitalist, industrial sector and rising savings and investment. For instance, in Colombia between 1953 and 1965 industrial output rose from 15·5 per cent of GDP to 19·0 per cent, while the gross investment ratio *declined* from 16·5 to 16·0 per cent. Similarly, in Guatemala between 1950–51 and 1962–63 the share of industry in GDP rose from twelve to fourteen per cent, while private savings as a per cent of net national product *declined* from 2·6 to 2·3. The same phenomenon has occurred in Brazil: during the period 1946–48 to 1958–60 industrial production increased from 21 to 34 per cent of GDP and the ratio of gross domestic savings to GNP declined from 16·4 to 16·0 per cent. Indeed it appears that the savings ratio in Brazil has remained roughly constant since at least the late 1930s.[40] The lack of a strong positive correlation between the degree of industrialization and the domestic savings ratio is not, of course, peculiar to Latin America; a similar lack of association can be found elsewhere. For example, in Turkey between 1954 and 1965 industrial output rose from thirteen to fifteen per cent of GDP, while the gross investment ratio—despite the availability of considerable amounts of foreign aid—fell from 14·5 per cent to 13·0 per cent of GDP. Thus once more the predictions of the theory are refuted by the facts.

In summary, dualistic models of development make an unhelpful division of the economy into a 'traditional' and a 'modern' sector. The assumptions of the theory regarding the characteristics of the 'traditional' agricultural sector are not credible and, indeed, can be shown to be erroneous by even the most casual empiricism. The predictions of the theory are likewise incorrect. It has been demonstrated that in not a few countries the growth of national income *per capita*, the level of rural wages, the expansion of agricultural output, the evolution of employment in industry and the behaviour of the

[38] 'Economic Development with Unlimited Supplies of Labour', *loc. cit.*

[39] 'Surplus Agricultural Labour and the Development of a Dual Economy', *loc. cit.*, p. 310.

[40] See N. H. Leff, 'Marginal Savings Rates in the Development Process: The Brazilian Experience', *Economic Journal*, September 1968.

aggregate savings and investment ratios have differed markedly from what the theory would lead one to expect. In almost every conceivable way the theory fails to conform to the reality of a great many underdeveloped countries.

In many respects this theory of growth and development is curiously static and a-historical. Models of the dual economy assume a given and constant subsistence wage rate, a given pool of disguised unemployment and unchanged, i.e. 'traditional', agrarian institutions. The real problem arises, however, when population growth rates exceed the capacity of the economy to adjust its institutions (e.g. land tenure), attitudes (e.g. toward birth control) and composition of output (e.g. the degree of industrialisation) so that real wages *fall*, seasonal unemployment in agriculture *increases* and the proportion of the labour force employed in large-scale industry *declines* or at best remains roughly constant. One cannot even begin to analyse these problems if the conceptual framework being used is one of static, unchanged, constant 'subsistence' incomes and 'traditional' institutions, values and modes of behaviour.

STAGE THEORIES

Economic history, and theories firmly based upon historical knowledge, would appear to be essential in understanding the nature of underdevelopment. Unfortunately, however, most of the theories which claim to view development in historical perspective begin by assuming that the underdeveloped countries are in a 'low-level equilibrium trap'.[41] This presumption, of course, largely precludes endogenous change, since the very essence of an equilibrium position is absence of movement.

A common procedure is to assume that all nations, rich and poor, were once equal, i.e. suffered from an equivalent degree of poverty and state of underdevelopment. The implications of Kuznets' findings that 'the present levels of *per capita* product in the underdeveloped countries are much lower than were those in the developed countries in their pre-industrialization phase'[42] have been totally ignored. Instead economists have argued from an assumption of equality when inequality obviously exists. Professor Leibenstein could not be more explicit. In defining 'the abstract problem' he says, 'We begin

[41] The phrase is taken from R. R. Nelson, 'A Theory of the Low-Level Equilibrium Trap in Underdeveloped Economies', *American Economic Review*, December 1956. Also see by the same author 'Growth Models and the Escape from the Low-Level Equilibrium Trap: The Case of Japan', *Economic Development and Cultural Change*, July 1960.

[42] S. Kuznets, *Economic Growth and Structure*, London 1966, p. 177.

with a set of economies (or countries), each 'enjoying' an equally *low* standard of living at the outset . . . . Over a relatively long period of time (say, a century or two) some of these countries increase their output per head considerably whereas others do not.'[43] This being so, the thing to do is determine how today's wealthy countries escaped the 'low-level equilibrium trap', and then apply the lessons to the backward countries which were left behind.

The most self-conscious attempt to do this is found in Rostow's book, *The Stages of Economic Growth: A Non-Communist Manifesto*. The terminology and analytical categories employed in this book, although severely criticized,[44] have permeated Western thinking on development problems.[45] The reasons for this have more to do with sociology than economics.

Rostow believes that all countries pass through five stages. The initial stage is called 'the traditional society' and its features are similar to those of the 'non-capitalist' sector of dual economy models. Next comes a 'preconditioning' stage, followed by the 'take-off', the 'drive to maturity' and, finally, an 'age of high mass-consumption'. How a nation gets from one stage to another is unclear, since all Rostow presents, in effect, is a series of snapshots which freeze the development process in five different moments of time. What is clear, however, is that the present 'traditional society' stage is the initial stage, and that development occurs essentially as a result of internal efforts which are largely unaffected by the workings of the wider international economy.

As Gunder Frank has stressed, Rostow's theory 'attributes a history to the developed countries but denies all history to the underdeveloped ones'.[46] Rostow neglects the past of the underdeveloped countries but confidently predicts a future for them similar to that of the wealthy nations. In this respect Rostow's views differ little from the Marxist doctrine that 'the most industrially advanced country presents the less advanced country with the image of its future'.

Marx and Rostow notwithstanding, it is exceedingly improbable

[43] H. Leibenstein, *op. cit.*, p. 4. Italics in the original.

[44] See, for example, P. A. Baran and E. Hobsbawm, 'The Stages of Economic Growth', *Kyklos*, 1961; S. Kuznets, 'Notes on the Take-Off', in W. W. Rostow, ed., *The Economics of Take-Off into Sustained Growth*, Macmillan, 1964.

[45] For instance, in presenting their dualistic model Ranis and Fei claim that Rostow's 'well-known intuitive notion has been chosen as our point of departure'. ('A Theory of Economic Growth', *loc. cit.*, p. 533).

[46] A. G. Frank, 'Sociology of Development and Under-Development of Sociology', *Catalyst*, Summer 1967, p. 37.

that one can gain an adequate understanding of present obstacles and future potential for development without examining how the underdeveloped nations came to be as they are. To classify these countries as 'traditional societies' begs the issue and implies either that the underdeveloped countries have no history or that it is unimportant.[47] No proof has yet been provided to substantiate either of these claims. Indeed it is clear that the underdeveloped countries do have a history and that it is important. Furthermore, evidence is gradually being accumulated that the expansion of Europe, commencing in the fifteenth century, had a profound impact on the societies and economies of the rest of the world. In other words, the history of the underdeveloped countries in the last five centuries is, in large part, the history of the consequences of European expansion. It is our tentative conclusion that the automatic functioning of the international economy which Europe dominated first created underdevelopment and then hindered efforts to escape from it. In summary, underdevelopment is a product of historical processes.

Historical research is gradually reconstructing the past of the underdeveloped countries for us. Enough is known to enable us to say with confidence that 'by the end of the sixteenth century . . . the agricultural economies of the spice islands, the domestic industries of large parts of India, the Arab trading-economy of the Indian Ocean and of the western Pacific, the native societies of West Africa and the way of life in the Caribbean islands and in the vast areas of the two vice-royalties of Spanish America [were] all deeply affected by the impact of Europeans . . . The results [of European expansion] on non-European societies were . . . sometimes immediate and overwhelming. . . .'[48]

The expansion of Europe throughout the world was an outcome of the competition among mercantilist-capitalist states for trading advantages. This competition was both peaceful and violent and its object was to obtain monopoly control of the most lucrative trading areas. In practice the quest for monopoly control led inevitably to the forceful acquisition of colonies, satellites, dependent territories and spheres of influence. But the initial impulse, from the time when

[47] A typical view is exemplified by Trevor-Roper's arrogant assertion that 'the history of the world, for the last five centuries, in so far as it has significance, has been European history'. (*The Rise of Christian Europe*, 1965, p. 11.)

[48] E. E. Rich, 'Preface', in E. E. Rich and C. H. Wilson, eds, *The Cambridge Economic History of Europe*, Vol. IV, *The Economy of Expanding Europe in the Sixteenth and Seventeenth Centuries*, Cambridge University Press, 1967, p. xiii. The contributions of Professor Rich to this volume, and especially the 'Preface', are brilliant.

the Portuguese first began to explore the Orient, was to dominate trade, not to gain territory; that came later. 'The object of Portuguese colonization was not the possession of the Indies themselves, but of the trade of the Indies. Their approach was based on a concept of a *mare clausum*, secured to them under papal authority, which should save them from the inroads of other Christian states, and on a system of forts and garrisons which should save them from native opposition.' The Portuguese had no wish to engage in production but 'merely to divert to their own sea-routes a trade which was based on a competent native economy. Their purpose was to make the king of Portugal the only merchant trading between India and Portugal'.[49]

Ironically, it was the combination of Europe's military superiority and her relative material poverty which shaped events in the early phase of European expansion. Western ascendancy was made possible by advanced military technology and it was made necessary by the inability of Europe to engage in trade on equal terms with the wealthy nations of the East. Asia had much that Europe wanted but Europe could offer almost nothing that was desired in Asia. As Professor Rich has said, 'the spice trade was conditioned by the fact that the Spice Islands wanted very little of the produce of Europe save firearms'.[50]

An historian of Indonesia notes that 'when the first Dutch merchants and sailors had come to the island world of the Indies, they had been amazed by the variety of its nature and civilization, and the more observant among them had recognized that southern and eastern Asia were far ahead of western Europe in riches as well as in commercial ability and mercantile skill'.[51] Similarly, an historian of the Middle East has written that 'when Islam was still expanding and receptive, the Christian West had little or nothing to offer, but rather flattered Islamic pride with the spectacle of a culture that was visibly and palpably inferior'.[52] Europe's subsequent ability to dominate the rest of the world depended not upon her cultural superiority or economic strength but upon two technological breakthroughs: the construction of large ocean-going sailing vessels and the development of gunpowder and naval cannons.[53] Indeed,

[49] E. E. Rich, 'Colonial Settlement and its Labour Problems', *ibid.*, p. 304.

[50] *Ibid.*, p. 368.

[51] B. H. M. Vlekke, *The Story of the Dutch East Indies*, Harvard University Press, 1946, p. 178.

[52] B. Lewis, *The Emergence of Modern Turkey*, Oxford University Press, 1961, p. 40.

[53] See C. M. Cipolla, *Guns and Sails in the Early Phase of European Expansion, 1400–1700*, Collins, 1965.

Europe owed a great technological debt to the rest of the world, and particularly to China. Without Chinese science the industrial revolution would have been impossible.[54] It was European advances in specific military techniques rather than general progress in the peaceful arts of civilization which enabled her to establish hegemony in Latin America, Asia and Africa.

In the early period of expansion, in fact, a large volume of trade between Europe and the rest of the world would have been impossible because of the European tendency to run a substantial balance of payments deficit. If Europe was to obtain the products from the East which were desired she either had to force down the price of oriental products or increase the demand for goods which Europe could supply. In practice she did both. The Dutch, for instance, exacted an annual tribute in spices; for other crops they enforced compulsory deliveries at favourable prices. The English destroyed the Indian textile industry and then proceeded to supply India with cotton goods from Great Britain. How Britain was to finance the imports of tea from China presented great problems, for as the Chinese emperor said to George III, 'our celestial empire possesses all things in prolific abundance' and, presumably, therefore, China had little need for English goods. This knotty problem was finally resolved by forcing opium on the Chinese and encouraging addiction. This created a large demand for the drug which the East India Company was able to supply from Bengal. The Chinese made many vain attempts to restrict the trade. Finally, Britain forced China to permit the trade and fought the Opium War of 1839–42—'a war that was precipitated by the Chinese government's effort to suppress a pernicious contraband trade in opium, concluded by the superior fire-power of British warships, and followed by humiliating treaties that gave Westerners special privileges in China'.[55]

It is still a matter of debate whether domination of the rest of the world was the vital ingredient in Europe's recipe for rapid economic growth. There is little doubt, however, that resources were transferred to the West, and especially to Great Britain, on a massive scale. British India had a large trading surplus with China and the rest of Asia. These surpluses, in turn, were siphoned off to England 'through the (politically established and maintained) Indian trading deficit with Britain, through the "Home Charges"—i.e. India's pay-

[54] See J. Needham and W. Ling, *Science and Civilization in China*, Vol. IV, Part II.

[55] J. K. Fairbank, E. O. Reischauer and A. M. Craig, *East Asia: The Modern Transformation*, George Allen and Unwin, 1965, p. 136.

ments for the privilege of being administered by Britain—and through the increasingly large interest-payments on the Indian Public Debt. Towards the end of the [nineteenth] century these items became increasingly important. Before the First World War "the key to Britain's whole payments pattern lay in India, financing as she probably did more than two fifths of Britain's trade deficits" '.[56]

Going back still further, the East India Company, according to Keynes, had its origin in privateering. 'Indeed, the booty brought back by Drake in the *Golden Hind* may fairly be considered the fountain and origin of British Foreign Investment. Elizabeth paid off out of the proceeds the whole of her foreign debt and invested a part of the balance (about £42,000) in the Levant Company; largely out of the profits of the Levant Company there was formed the East India Company, the profits of which during the seventeenth and eighteenth centuries were the main foundation of England's foreign connections; and so on. In view of this, the following calculation may amuse the curious. At the present time (in round figures) our foreign investments probably yield us about 6½ per cent net after allowing for losses, of which we reinvest abroad about half—say 3¼ per cent. If this is, on the average, a fair sample of what has been going on since 1580, the £42,000 invested by Elizabeth out of Drake's booty in 1580 would have accumulated by 1930 to approximately the actual aggregate of our present foreign investments, namely £4,200,000,000—or, say, 100,000 times greater than the original investment. We can, indeed, check the accuracy of this assumed rate of accumulation about 120 years later. For at the end of the seventeenth century the three great trading companies—the East India Company, the Royal African and the Hudson's Bay—which constituted the bulk of the country's foreign investment, had a capital of about £2,150,000; and if we take £2,500,000 for our aggregate foreign investments at that date, that is of the order of magnitude to which £42,000 would grow at 3¼ per cent in 120 years'.[57]

The Keynesian calculation is indeed amusing (particularly to an Englishman!) and well illustrates what Rostow calls 'the march of compound interest'.[58] We cannot tarry here, however, for we are concerned not with whether European expansion enriched the West, but with whether it impoverished the rest of the world. It is conceivable

[56] E. J. Hobsbawm, *Industry and Empire*, Weidenfeld and Nicolson, 1968, p. 123, citing S. B. Saul, *Studies in British Overseas Trade 1870–1914*.

[57] J. M. Keynes, *A Treatise on Money*, Vol. II, *The Applied Theory of Money*, Macmillan, 1930, pp. 156–7.

[58] *Op. cit.*, p. 6.

that the benefits to Europe of its hegemony were slight and accrued in the form of temporarily increased consumption (rather than greater investment and growth), while the costs of her dominance were heavy and fell primarily upon the dependent countries. It is to this final question that we now turn.

FRAGMENTS OF HISTORY

The concept of 'underdevelopment' as it is used in this book is all-inclusive. It refers to a society's political organization, economic characteristics and social institutions. Poverty is neither a synonym for underdevelopment nor a cause of underdevelopment; it is only symptomatic of a more general problem. Poverty, in other words, forms part of a culture. Oscar Lewis had the following to say about this culture: 'The culture of poverty is both an adaptation and a reaction of the poor to their marginal position in a class-stratified, high individuated, capitalistic society. It represents an effort to cope with feelings of hopelessness and despair. . . . Most frequently the culture of poverty develops when a stratified social and economic system is breaking down or is being replaced by another. . . . Often it results from imperial conquest in which the native social and economic structure is smashed and the natives are maintained in a servile colonial status, sometimes for many generations'.[59]

As Lewis is the first to recognize, however, the culture of poverty is not identical in all settings; the slums of Puerto Rico produce a different culture from those of Mexico City;[60] the culture of poverty varies from place to place and from one era to another. The culture both shapes and is shaped by a people's history. It is for this reason that the differences between the developed and the underdeveloped countries cannot be explained exclusively in statistical terms; the two types of countries differ qualitatively as well as quantitatively. For similar reasons, it is almost certainly incorrect and misleading to assume that the circumstances of today's underdeveloped countries were always the same. Yet this is the view that at present prevails. Nurkse's notion of the 'vicious circle of poverty'—the proposition that 'a country is poor because it is poor', and presumably always has been—expresses the conventional doctrine perfectly.[61] As an alternative approach one might advance the hypothesis that the well-being of today's poor countries was not always so low and that their

[59] Oscar Lewis, *La Vida*, Secker and Warburg, London, 1967, p. xli.

[60] See Oscar Lewis, *The Children of Sanchez*, Secker and Warburg, 1961.

[61] R. Nurkse, *Problems of Capital Formation in Underdeveloped Countries*, Oxford University Press, p. 4.

descent into underdevelopment did not occur independently of what was happening in the rest of the world.

It is our belief that underdeveloped countries as we observe them today are a product of historical forces, especially of those forces released by European expansion and world ascendancy. Thus they are a relatively recent phenomenon. Europe did not 'discover' the underdeveloped countries; on the contrary, she created them. In many cases, in fact, the societies with which Europe came into contact were sophisticated, cultured and wealthy.

This is well illustrated by the case of Indonesia, an archipelago which today includes about half of the inhabitants of South-East Asia and the region which formerly acted as a magnet to Western traders and precipitated European expansion. At the beginning of the sixteenth century Indonesia was a prosperous region. 'Local emporia were the equal of anything Europe had to offer: indeed Malacca was at that time regarded by Western visitors as the greatest port for international commerce in the world, clearing annually more shipping than any other.'[62] The Dutch, operating through the Netherlands' United East India Company, aimed first to establish a monopoly of trade with the region. This aim was accomplished by 1641. They next established a monopsony over the purchases of the output of the islands. Finally, in the eighteenth century, the Dutch established a system of forced deliveries, forced cultivation and even the legal obligation to grow specific commercial crops on peasant holdings. Specialization was not dictated by the market but by the Company. As a consequence of this so-called Culture System 'so little time was left to the Javanese for the cultivation of food crops that serious famines occurred in the eighteen-forties. The fertile island had been transformed into a vast Dutch plantation, or, from the point of view of the people, a forced labour camp'.[63]

Agriculture was not the only sector that was adversely affected. The Dutch systematically discouraged and prevented local enterprise outside agriculture, and even brought in Chinese as ubiquitous middlemen. Java's indigenous commercial and industrial activities were utterly destroyed: ship building, iron-working, brass and copper founding all disappeared; weaving and peasant handicrafts declined; the merchant marine vanished and the merchants devoted themselves to piracy.

By the beginning of the present century the Indonesian economy was in a state of crisis and the Dutch government announced its

[62] M. Caldwell, *Indonesia*, Oxford University Press, 1968, p. 39.
[63] *Ibid.*, p. 47.

intention in 1901 to 'enquire into the diminishing welfare of the people of Java'. Some indication of the extent to which the well-being of the people had declined is provided by Mr Caldwell's figures:

Table 3

AVERAGE ANNUAL RICE CONSUMPTION PER HEAD IN JAVA AND MADURA

Period	*Quantity* (*kilogrammes*)
1856–70	114·0
1881–90	105·5
1891–1900	100·6
1936–40	89·0
1960	81·4

Source: M. Caldwell, *op. cit.*, p. 21.

Indonesia's experience was not unique. Indeed, President Roosevelt's comment to Lord Halifax in January 1944 that the French had possessed Indochina '. . . for nearly 100 years, and the people were worse off than they were at the beginning' is applicable to Asia as a whole. In some cases the destruction of the indigenous society was largely inadvertent. The decimation of the population of the South Pacific islands through the introduction of alien diseases is an example of this.[64] In other cases the destruction of the native economy and its institutions was deliberate. A second great example of this is India.

As late as the early seventeenth-century India was more advanced economically than Europe. She had a fairly large manufacturing sector which produced mostly luxury goods—including gold and silver objects, plus glassware, paper, iron products and ships. Many of these items as well as cotton cloth, silk, indigo and saltpetre were exported to the West for payment in bullion.[65] The decline of India's industry was due to a combination of several factors: technical progress in Europe associated with the industrial revolution, domination of the East India Company and the imposition of the free trade doctrine under unequal conditions by the British. After 1833 the process of de-industrialization was accelerated and emphasis was placed on developing cash crop agriculture for export. Industrial decay was complete by the 1880s.

[64] See A. Moorehead, *The Fatal Impact*, Hamish Hamilton, 1966, Part I.

[65] S. C. Kuchhal, *The Industrial Economy of India*, Chaitanya Publishing House, 1965, p. 64.

Parallel to the destruction of the manufacturing sector, agricultural institutions were profoundly altered and the economic wellbeing of rural inhabitants declined. Throughout the nineteenth century the proportion of the total population dependent upon agriculture increased, and the proportion of the rural population composed of agricultural labourers also increased. Data from the Madras Presidency of South India indicate that the real wages of agricultural labourers (measured in *seers* of common rice) declined sharply even as late as the last quarter of the last century. In only one of the seven districts for which data are available did real wages actually rise; in the others they declined from 13 to 48 per cent.

Table 4

CHANGE IN REAL WAGES OF AGRICULTURAL LABOUR IN SEVEN DISTRICTS OF SOUTH INDIA, AVERAGE 1873–75 TO AVERAGE 1898–1900

	per cent
Ganjam	−43
Vizagapatam	−48
Bellary	−20
Tanjore	+29
Tinnevelly	−40
Salem	−13
Coimbatore	−39

Source: Dharma Kumar, *Land and Caste in South India*, Cambridge University Press, 1965, p. 164.

Conditions in the rest of India were roughly comparable. René Dumont summarizes the experience of Bengal as follows: 'On 22 March 1793 Lord Cornwallis and the East India Company proclaimed that *zamindars* and *talukhars* (the men who had been charged with the collection of tribute) would henceforth be considered as permanent and irrevocable owners of the lands on which they had gathered taxes. This proclamation had far-reaching consequences. Of course, it is easy to see that the East India Company regarded it both as an improved way of obtaining a better return of tributes, and also as an easy means of making firm allies. But they never realized that, in depriving the peasant of his traditional and permanent right to occupy the land, they were making him, throughout the greatest part of India, a slave of new owners; and that exploitation of the peasant now took the place of exploitation of resources. Rural societies were not only compelled to pay taxes, but also rents which demographic development soon made outrageous; some peasants took to running away. A new law gave the *zamindars* the right to catch them, and this completed the dismemberment of traditional rural society. On the one hand great landowners; serfs on the other; the former with no incentive to improve the land; the latter with no means to do so'.[66]

The conversion of tax collectors into landlords, the emphasis on production of cash crops for export, and the population explosion which began at the end of the nineteenth century were jointly responsible for the final disaster. The mass of the people were reduced to a subsistence income which hovered precariously above the famine level. Using 1900–01 as an index base of 100, agricultural production per capita had declined to 72 a half century later, while production of food *per capita* had plunged to the miserably low figure of 58.[67]

None of the preceding discussion should be taken to imply that all of the underdeveloped countries were once wealthy societies and advanced civilizations. Some of the peoples with whom the Europeans came into contact were, of course, relatively primitive. But nearly all of the people encountered in today's underdeveloped areas were members of viable societies which could satisfy the economic needs of the community. Yet these societies were shattered when they came into contact with an expanding Europe. The manner in which the indigenous societies were destroyed varied from one region to another and depended upon the precise form taken by European penetration and the wealth, structure and resilience of the native civilization. Although the method of destruction varied, the outcome was always the same: a decline in the welfare of the subjugated people. Writing about Africa, Professor Frankel notes that attempts at modernization under colonialism are 'in greater or lesser degree accompanied by increasingly rapid disintegration of the indigenous economic and social structure. However primitive those indigenous institutions may now appear in Western eyes, they did in fact provide the individuals composing the indigenous society with that sense of psychological and economic security without which life loses its meaning'.[68]

[66] R. Dumont, *Lands Alive*, Merlin Press, 1965, p. 139.

[67] See K. Mukerji, *Levels of Economic Activity and Public Expenditure in India*, Asia Publishing House, 1965.

[68] S. H. Frankel, *The Economic Impact on Underdeveloped Societies*, Basil Blackwell, 1953, p. 134.

Although our knowledge of African history is rudimentary, it is perhaps correct to say that no continent has felt the impact of European expansion more thoroughly than Africa. The introduction, especially by the Portuguese, of large-scale trading in slaves during the sixteenth century completely disrupted West Africa from Guinea to Angola.[69] Slavery created chaos in vast areas of the continent. The population declined; wars among formerly peaceful tribes were incited; the native economy fell into decay; and the social organization of the community and the authority of the chief frequently were corrupted. The entire way of life in Africa was altered. 'The increased demand for slaves arising from the plantation owners of North and South America in the seventeenth and eighteenth centuries was responsible for depopulating large parts of Africa, and for degrading what had once been settled agricultural peoples back to long-fallow agriculture or nomadism'.[70]

The slaughter of the indigenous people and the depopulation of the land did not cease with the end of slavery, however. In 1919 the Belgian Commission for the Protection of the Native estimated that the number of inhabitants of the Congo had declined by as much as 50 per cent since the beginning of occupation forty years earlier. In South-West Africa during the German-Herero War of 1904 General von Trotha, after the campaign was over, issued his notorious Extermination Order which required every Herero man, woman and child to be killed.[71] As a result of this the tribe was reduced from 80,000 to 15,000, and today it has regained only half of its former strength.

As pervasive as slavery and indiscriminate slaughter may have been, they can hardly be considered the typical pattern of European penetration in Africa. One must also consider the more 'normal' economic activities of colonization and mineral extraction. One cannot, of course, accurately describe in a few paragraphs all the forms which colonialism adopted in North, East and Southern Africa, but it is possible to reconstruct a simplified scheme of the effects of European activity upon the indigenous society.

[69] See J. Duffy, *Portuguese Africa*, Harvard University Press, 1959, especially Ch. VI. See also the well-known study by E. Williams, *Capitalism and Slavery*, University of North Carolina Press, 1944. J. Pope-Hennessy, *Sins of the Fathers: A Study of the Atlantic Slave Trade, 1441–1807*, 1967, is a lively popular account.

Slave raiding in Eastern and Central Africa had been introduced earlier by Arab traders operating out of Zanzibar and Khartoum. This naturally disturbed the native economy and society, but the effects were insignificant in comparison with the devastation created by European and American slaving expeditions.

[70] Colin Clark, *Population Growth and Land Use*, Macmillan, 1967, p. 136.

[71] See R. First, *South-West Africa*, Penguin, 1963, pp. 69–83.

The process began with the acquisition of all the good land, mineral deposits and water resources by the colonialists. Excluding West Africa, this was nearly a universal phenomenon, and was not confined to the acknowledged cases of white settlers in Kenya, Algeria and the Republic of South Africa, but was also prevalent in less prominent places. For instance, the Bechuana tribes of Botswana were continually forced to give up their most productive lands in the south and northwest in order to avoid becoming a colony and to maintain their status as the Bechuanaland Protectorate.[72] In Liberia the descendants of freed slaves (Americo-Liberians) have installed themselves as aristocratic absentee landlords of rubber farms, have required the indigenous people to supply one fourth of the labour supply gratis, and pay the remainder four cents an hour or less.[73] The mandate territory of South West Africa is a classical example of Europeans monopolizing the land. 'Whites, though only one in seven of the total population, enjoy the exclusive use of two-thirds of the land.'[74]

Having lost the best lands, the indigenous population was then confined to the less desirable and more remote areas—the 'bush', Reserves, the veld or Bantustans. The high population densities led inevitably to increased erosion, declining yields of food crops in native areas and falling consumption levels. Colonialism in Africa—like that in Latin America, as we shall soon see—led to underemployment both of land (in the European areas) and labour (in the

Table 5

PER CAPITA OUTPUT OF INDIGENOUS AGRICULTURE IN ALGERIA

	Cereals (kilos)	Cattle (head)	Sheep (head)
1863	1000	n.a.	4·5
1911	377	0·2	1·5
1938	231	0·1	0·8
1954	202	0·1	0·7

Source: R. Murray and T. Wengraf, 'The Algerian Revolution', *New Left Review*, No. 22, p. 32, who cite A. Gorz, "*Gaullisme et neo-colonialisme*', *Temps Modernes*, March 1961.

[72] E. S. Munger, *Bechuanaland*, Oxford University Press, 1965, Ch. II.

[73] G. Dalton, 'History, Politics, and Economic Development in Liberia', *Journal of Economic History*, December 1965.

[74] R. First, *op. cit.*, p. 142.

African areas). *Per capita* food consumption, at least in some cases, has fallen over a considerable period of time. For example, food consumption in Algeria was perhaps between five and six times higher in 1863 than it was in 1954.

It was not sufficient, however, simply to dispossess the natives of their land and confine them to Reserves. The colonial economy—particularly the mines—also required cheap manpower; the Africans had to be compelled to emigrate and work for the Europeans. In some cases, e.g. in the Belgian and Portuguese colonies, the authorities relied to a great extent on forced labour. In most of the other colonies, however, a more subtle device was used—fiscal policy. A high tax, payable in money, was imposed on the natives. This forced them to enter a monopsonistic labour market and work for the white men at extremely low wages in order (i) to pay their taxes and (ii) to supplement the declining income obtainable from indigenous agriculture. Positive inducements in the form of incentive goods also were occasionally provided. Often this was unnecessary, however. A common technique, as in Basutoland, was to assign the responsibility for collecting taxes to the chief and allowing him to take a rake-off. In this way the authority of the chief was used to favour the ambitions of the colonialists rather than the interests of his own people. The system of colonialism and indirect rule was designed to generate abundant supplies of cheap unskilled labour for Europeans who monopolized all other resources. The material wellbeing of the African was systematically lowered and his institutions were intentionally destroyed. It was this process of impoverishment and growing degradation which contributed to the urgent demands for independence in the late 1940s. By this time Africa and the other underdeveloped countries had gone through a lengthy period of growing misery which culminated in the collapse of primary commodity prices in the 1920s, the world depression of the 1930s, and the Second World War of the first half of the 1940s. The crisis of colonialism was not exclusively or even primarily a political crisis; its roots lay in the inability of the colonial system to generate economic progress and distribute it equitably.[75]

Even this rather superficial discussion of conditions in Africa and Asia should give us a broader perspective from which to consider the historical origins of underdevelopment in Latin America, the region which embraces the Spanish American nations with which we shall be directly concerned in later chapters. In general, colonialism in Latin America, as in the rest of the world, was a catastrophe for

[75] See B. Davidson, *Which Way Africa?*, Penguin, 1964, Ch. 6.

the indigenous people. In the areas of more primitive civilization the population virtually disappeared within less than thirty years. In the areas of advanced civilization the people were completely subjugated.

Spanish penetration of Latin America began in the Caribbean area. There they encountered Arawak, Carib and Cueva tribes with large populations tilling the soil in permanent clearings and on *conucos*. The native culture in the West Indies and on the Isthmus was not as advanced as some other civilizations, but the tribal societies were well organized and the economy was perhaps as productive as that of Indonesia. Yet within a generation the indigenous society and economy had been ruined and the native population had virtually disappeared.[76]

The Spaniards gained control over the natives by breaking their political structure. The Chiefs were liquidated and the rest of the community were allocated to individual claimants. These allocations were originally called *repartimientos* and subsequently formed the basis of the *encomienda* system. These colonial institutions, in turn, were the origin of the latifundia system, under which individual rights to labour services were transformed to include the land as well. One of the features of the *repartimientos* was that the number of natives allocated to a Spaniard depended upon how much work he could extract from them, i.e. originally, how much gold for export he could get them to produce. In this way strong incentives to exploit labour were created.

The combination of brutality, slaughter, high tribute, slavery, forced labour for gold mining, destruction of the social framework, malnutrition,[77] disease and suicide led to the extinction of the indigenous population. 'It has been reckoned that at the approach of the Spaniards, in 1492, total Carib population in Hispaniola was about 300,000. By 1508 it was reduced to about 60,000. A great decline had brought it to about 14,000 by 1514, as serious settlement began; and by 1548 it had reached a figure which indicated virtual extermination, about 500.'[78] The population of the other islands

[76] See C. O. Sauer, *The Early Spanish Main*, University of California Press, 1966, especially chapters III and VII.

[77] There was never a deficiency of cassava bread and sweet potatoes on the islands. Malnutrition occurred after the Spaniards suppressed native fishing and hunting and the supply of protein and fat declined.

[78] E. E. Rich, 'Colonial Settlement and its Labour Problems', *loc. cit.*, p. 319. The author adds that 'European diseases had played their parts in this decimation of the Carib population, but the main cause was without doubt a passive revulsion from the changes which white occupation brought.'

declined even more rapidly. The Bahamas lost their population first. Puerto Rico was decimated in little more than a decade, and Cuba followed soon after. By 1519 Jamaica was almost uninhabited. Those who survived were a pitiable lot. 'A well-structured and adjusted native society had become a formless proletariat in alien servitude. . . .'[79]

As the population declined the *conucos* on the islands were abandoned and the terrain became rangeland for cattle and pigs; in Central America the continuous savanna reverted to a tropical rain forest. The Spaniards responded to the labour shortage by introducing extensive grazing on their estates. The few natives who managed to escape fled to the jungle and adopted the slash-and-burn shifting agriculture that can still be observed today.

A similar story may someday be told, perhaps, of the sparsely settled regions of the Amazon basin. It is usually assumed that this region was inhabited by extremely primitive people: this assumption, however, may well turn out to be incorrect. The inhabitants of this area may once have had a more advanced civilization and a higher standard of living than is currently believed. A noted anthropologist who has had considerable research experience in Brazil, Claude Lévi-Strauss, is too cautious to advance a positive hypothesis, but the question he poses is worth pondering. 'Is it not also possible to see them [the tribes in Brazil] as a regressive people, that is, one that descended from a higher level of material life and social organization and retained one trait or another as a vestige of former conditions?'[80]

We do not know what the answer to his question is as regards Brazil, but in the two cases of Mexico and the Inca Empire the answer is clearly 'yes'. Space does not permit us to recount the downfall of the Aztecs. Let us only note that the native population of Mexico was decimated. From about 13 million at the time of the Spanish Conquest, the population had declined to about 2 million by the end of the sixteenth century.

In the Inca Empire, which covered a very large portion of Western South America, the impact of the Spanish was not quite so fatal, yet it is still true that one of the greatest tragedies in Latin America was the destruction of this civilization. The Spanish Conquest of Peru was accompanied by profound social, institutional and demographic changes. The wars, the epidemics and the fierce exploitation of the

[79] C. O. Sauer, *op. cit.*, p. 204.

[80] Claude Lévi-Strauss, *Structural Anthropology*, Anchor Books, 1967, p. 101. Also see by the same author, *Tristes Tropiques*, 1958.

Indians reduced the indigenous population by a half to two-thirds.[81] It was only towards the end of the nineteenth century that the Indian population began to increase again, and it is now estimated that this population only slightly exceeds the number of inhabitants of the Inca Empire. The catastrophic decline in population was accompanied by the utter ruination of the Andean civilization. Cities vanished; the communal customs of the Inca became an historical curiosity; terraced hillsides were abandoned; agricultural productivity declined. The survivors of the conquest became a miserable, starving, diseased and disorganized mass of humanity. In short, they became an underdeveloped people.[82]

The new civilization constructed from the debris of the earlier indigenous society was markedly different. The colonizing Spaniards and their descendants enslaved what remained of the indigenous population. Indians were sent to the mines by the thousands to extract the mineral wealth of the continent. Following the precedent established in the Caribbean, the best lands were appropriated and huge estates were distributed to the favoured few. The great mass of the underprivileged, on the other hand, were pushed on to the mountain slopes where they attempted to eke out a living on small plots. In this way the distinctive economic system of Spanish America—the latifundia-minifundia complex—was created.

The essential feature of the new economic system was the monopolization of land. This by itself was sufficiently important to shape the social and political relationships of the colonial civilization, since in a predominantly agricultural economy one's livelihood depends almost entirely upon access to land. Exploitation did not stop here, however. Water rights were tightly controlled by the large landowners; the majority of the population had very little access to credit; rural education was practically non-existent. Thus the latifundium acquired a monopoly of the major factors of production—land, capital, water and technology, and its position as virtually the

[81] In Latin America as a whole Colin Clark estimates that the population declined from 40 million in 1500 to 12 million in 1650. (*Population Growth and Land Use*, p. 64.)

In North America the indigenous population was not very large, but the Indian was destroyed never-the-less. A former US Commissioner of Indian Affairs has described in detail how 'a policy at first implicit and sporadic, then explicit, elaborately rationalized and complexly implemented, of the extermination of Indian societies and of every Indian trait, of the eventual liquidation of Indians, became the formalized policy, law and practice'. (John Collier, *Indians of the Americas*, Mentor, 1948, p. 103.)

[82] The classic study of this process is W. H. Prescott, *The Conquest of Peru.*

only large employer gave it a strong monopsonistic position in the labour market as well. The economic power of the minifundium was nil; its role in the system was to provide an abundant supply of cheap, unskilled labour.

Low productivity and an unequal distribution of income were inevitable characteristics of the new social and economic system. The universal syndrome of the latifundia-minifundia complex was the continuous pressure upon the Indians to move to poorer lands, the consequent accelerated erosion of the mountain slopes, falling yields of food crops on the subsistence plots, and a decline in consumption standards of the mass of the population. In contrast to the intensive agriculture of the minifundium and its declining productivity, the latifundium adopted highly labour extensive techniques of production and the large landowners were able to prosper at the expense of the rest of the community. Thus it was the social and political systems imposed by the colonists, in combination with the demographic changes which followed the Conquest, which were responsible for creating underdevelopment in Spanish America. One cannot explain the poverty of the region today without referring to the region's history.

CONCLUSIONS

Underdevelopment as it is encountered today in Spanish America and elsewhere is a product of history. It is not the primeval condition of man, nor is it merely a way of describing the economic status of a 'traditional' society. Underdevelopment is part of a process; indeed, it is part of the same process which produced development. Thus an interpretation of underdevelopment must begin with a study of the past. It is only from an examination of the forces of history—i.e. of the historical uses of power, both political and economic—that one may obtain an insight into the origin of underdevelopment.

The study of the uses of power in the past must be complemented by an analysis of the distribution of power in the present. The opportunities for development are conditioned by the functioning of the world economy in which the underdeveloped countries find themselves. There are some international economic forces which obviously tend to stimulate development, but there are many other forces which perpetuate inequalities and tend to retard development. It is argued below, especially in the concluding chapter,[83] that the transfer

[83] Also see P. P. Streeten, 'The Frontiers of Development Studies: Some Issues of Development Policy', *Journal of Development Studies*, October 1967, pp. 2–8.

of ideas, knowledge, factors of production and commodities may all increase rather than decrease the obstacles to development.

The internal barriers to development—e.g. inappropriate institutions, attitudes and values—are as important as the external obstacles. Moreover, as we argue in Chapter I, the types of barriers one finds, and their strength, frequently are related to the way economic and political power are distributed within the country. The concentration of purchasing power and the instruments of legitimate political force in a few hands, and the use to which this force is put, inevitably affect a country's aggregate economic performance and the welfare of its inhabitants.

Broadly speaking, the object of development policy is to turn historical constants into variables. Occasionally this can be achieved merely by changing the pattern of expenditure and the composition of investment. For example, the government might spend more of its revenues on rural education and less on central administration; investment in military installations might be reduced and expenditures on directly productive activities increased. In many cases, however, policy in underdeveloped countries cannot be concerned exclusively with allocating resources in the usual sense; it must also be concerned with creating new institutions and reforming existing ones. The major purpose of development planning, in fact, is to undertake the required structural transformation of a country in a conscious, explicit, orderly and rational manner.

The essence of development is institutional reform.[84] This process of institutional reform can act as an independent variable stimulating growth, e.g., an educational reform can stimulate growth by increasing the supply of relevant skills and improving the quality of the labour force. Alternatively, institutional reforms may be a prerequisite to development, e.g., large-scale investment in some minifundia zones may be virtually impossible unless fragmented land

[84] In most of the literature on growth and development institutional reforms are ignored; in other cases reforms are treated in a purely formal way. For example, Mrs Adelman writes her production function in the form $Y = f(K_t, \ldots, U_t)$, where K_t is the stock of capital at time t and U_t is the 'socio-cultural environment'. By differencing the production function one can determine the effect of institutional reform,

$$\sum_{g=l}^{w} \frac{\Delta Y}{\Delta Uj} \cdot \frac{\Delta Uj}{\Delta t}$$

This last expression, however, increases our knowledge of the role of institutional reforms by precisely zero. It tells us that if reforms lead to increased output, they lead to increased output; and if they don't, they don't. A more trivial result is hard to imagine. (See I. Adelman, *Theories of Economic Growth and Development*, Stanford University Press, 1961, Ch. 2.)

holdings are consolidated. Most important, institutional reforms may be complementary to other development policies and increase their effectiveness, e.g. a reorganization of the government's administrative machinery may be essential if development policies are to be properly formulated and implemented.

The three reforms we have mentioned—of the educational system, land tenure and public administration—are just a few of the many that are required. Furthermore, most of these reforms are linked to others. For example, the government administration cannot be improved unless the educational system is altered; education reform is contingent upon increased tax revenues; tax reform is impossible unless the political power of the wealthy is reduced and this, in turn, requires a land reform. The outcome of such a series of reforms is little short of a revolution. This is what Paul Baran meant when he said that 'economic development has historically always meant a far-reaching transformation of society's economic, social, and political structure'.[85]

[85] P. Baran, *The Political Economy of Growth*, Monthly Review Press, 1957, p. 3.

UNDERDEVELOPMENT IN SPANISH AMERICA

CHAPTER I

Spanish America: The Social and Economic Structure

This study is concerned with the social progress and economic development of the Spanish speaking countries of South America, namely, Argentina, Bolivia, Chile, Colombia, Ecuador, Paraguay, Peru, Uruguay and Venezuela. In spite of their diversity these nine countries seem to exhibit common features that go beyond the mere fact that Spanish is the official language in each of them. Their history, their social structure, the pattern of their international trade and the economic relations they have with the rest of the world are such that a number of generalizations can be made about the way their economies have behaved in the past as well as about what measures are necessary to improve performance in the future.

1 INTRODUCTION

One of the prominent features of Spanish America is the fragmented labour and capital markets and the defective functioning of the price system. This in turn has led to a splintering of economic activity into poorly articulated or imperfectly integrated sectors. The relatively slight volume of trade in intermediate goods between sectors of the economy is a reflection of this poor articulation. Equally important, each of the separate sectors is characterized by distinctive institutions—for example, the way in which assets are held, markets are organized or information conveyed. The response of these institutions to similar stimuli, e.g. a price signal, is quite different, and as a result, a given response can only be achieved by introducing a series of discriminatory policy measures in each sector. For example, the injection of capital in the Peruvian highlands may not be sufficient to increase output, since knowledge of improved agricultural techniques also is lacking. On the other hand, additional investment may not be necessary to increase food production in Chile's Central Valley, since all that may be required is a reform of legislation regarding water rights. Finally, increased investment may be both necessary and sufficient to increase energy production in Argentina. Similar examples can easily be found within as well as between countries.

In general, our Spanish American nations are split into two major sectors: the rural, agricultural sector, which may include anywhere from 32 to 65 per cent of the population, and the equally as large urban areas.

Table I: 1

URBAN POPULATION IN SPANISH AMERICA
(per cent)

Argentina	68·0	Paraguay	35·4
Bolivia	35·0	Peru	47·5
Chile	68·2	Uruguay	66·0
Colombia	52·0	Venezuela	62·5
Ecuador	36·0		

Source: UN, *Monthly Bulletin of Statistics;* ECLA, *Statistical Bulletin for Latin America.*

The agricultural sector is sub-divided into minifundia—which are largely subsistence farms—and the associated latifundia and plantations. The latter may be owned either by domestic or foreign interests, as for example in Paraguay. The urban sector usually is sub-divided into petty services and government (which together include the urban disguised unemployed) and the modern manufacturing and extractive industries, plus the associated transport, banking and financial services. In many countries, notably Uruguay, the state bureaucracy is the most important source of low wage, low productivity employment. The modern urban sector frequently includes foreign as well as domestic businesses, and in the relatively more industrialized economies may include some autonomous public-sector corporations.

The average annual rates of growth of GNP *per capita* in Latin America as a whole are below those of most other regions of the world. Moreover, Latin America's average growth rate appears to have been falling steadily since 1950. Within the nine Spanish American countries, Uruguay and Bolivia probably have moved backwards; the rate of growth in Venezuela and Ecuador has declined; Peru has made rapid progress and the others have grown at more modest rates. Some indication of the aggregate growth performance of the region is provided by the data in Table I:2.

Table I: 2

ANNUAL RATE OF GROWTH OF GNP *PER CAPITA*
(per cent)

	1950–55	*1955–60*	*1960–65*	*1966*
Argentina	1·4	0·9	1·3	−2·6
Bolivia	−1·5	−2·4	2·0	3·0
Chile	0·7	1·9	1·6	4·2
Colombia	2·6	1·1	1·5	2·1
Ecuador	2·3	1·4	1·3	1·6
Paraguay	0·7	0·2	1·6	1·4
Peru	3·7	2·0	3·7	3·0
Uruguay	n.a.	−1·4	−0·9	−0·5
Venezuela	5·1	3·2	1·6	−0·5
Latin America	2·3	2·0	1·5	1·3

Source: Agency for International Development, Statistics and Reports Division.

The above information should be interpreted with great care, however, as the data on national income are very unreliable. In the first place, it is difficult to talk of 'the' recorded rate of growth. There are usually several sources of data on growth rates—the Central Bank, the Planning Commission, the Central Statistical Office, the Economic Commission for Latin America, IBRD, AID, etc.—and these frequently are contradictory. In Peru, for example, the Central Bank and the planning institute publish separate series of national product; not only do the growth rates differ but on occasion the direction of change of GNP also differs.

Next, even if one does not have a plethora of contradictory information, a time series of GNP expressed in constant prices is likely to be quite unreliable. Inflation has been very rapid in almost all of the Spanish American countries. For example, between 1958 and 1964 prices rose 70 per cent in Bolivia, 90 per cent in Colombia, 300 per cent in Chile, 350 per cent in Uruguay and about 500 per cent in Argentina.[1] Under these conditions converting current prices to constant prices will be extremely difficult and small errors in the deflating procedure will be associated with large errors in the estimate of real growth rates. The difficulties will be accentuated if price controls are introduced in an attempt to reduce inflation; black markets will flourish and quality will deteriorate.[2]

[1] IMF, *International Financial Statistics.*

[2] For example, bakers avoid price controls on bread by reducing the size of the loaf and using lower quality flour; milk vendors avoid controls by diluting milk with water scooped out of the gutter.

After a few years of this the price indexes will become almost meaningless and deflated national income data will be impossible to interpret.[3]

None of this implies that GNP *per capita* has declined in Spanish America. It is quite likely, however, that as far as the majority of the population are concerned growth has been largely illusory. Moreover, it is quite possible that whatever growth has been achieved cannot be sustained.

Where growth has occurred it frequently has been due to a rapid expansion of foreign demand for primary exports,[4] e.g. petroleum and iron ore in Venezuela in the early 1950s. In many cases, a rise in foreign demand leads mainly to a rise in the profits of foreign owned extractive industries (and hence leak abroad) or a rise in the incomes of plantation owners. In either event, the majority of the population benefits relatively little.[5]

The second major growth sector has been in manufacturers. Much of the industrialization in Spanish America has occurred behind very high tariff walls. Very little growth in manufacturing can be attributed to the spontaneous initiative of native entrepreneurs; on the contrary, much of the industrial growth has been due either to foreign

[3] For instance, how is one to interpret the Chilean data which show negative household savings persisting for decades? The most recent data are as follows:

CHILE: Savings of Persons and Non-Profit Institutions, 1960–1966
(millions of *escudos*)

Year	Savings	Year	Savings
1960:	– 294	1964:	– 416
1961:	– 124	1965:	– 284
1962:	– 138	1966:	– 486
1963:	– 455		

Source: ODEPLAN, *Cuentas Nacionales de Chile 1960–1966*, Santiago, August 1967, Table 5.

[4] This is not expected to continue. 'Projections prepared by the Secretariats of two United Nations agencies and of the General Agreement on Tariffs and Trade (GATT) indicates that Latin American exports to the outside world over the next 20 years will increase at annual rates ranging from less than 2 per cent per annum to a high of nearly 4 per cent.' (United States Congress, Committee on Foreign Relations of the United States Senate, *Problems of Latin American Economic Development*, Feb., 1960, p. 9.) These export projections should be compared with the 6.5 per cent annual rate of growth of imports of the region. Also see Chapter II below.

[5] For a detailed analysis of these problems in a somewhat different context see R. E. Baldwin, *Economic Development and Export Growth*, 1966.

investment or to government financed and administered corporations in transport, energy, extractive or basic manufacturing activities. The manufacturing investments sponsored by private domestic entrepreneurs usually are located either in small consumer goods industries—for example, textiles, beer, leather goods, and furniture—or are satellite factories to the large foreign and government managed enterprises. Even these private manufacturing investments frequently were undertaken by immigrants[6]—who no longer are attracted to the region in large numbers.[7] This, in combination with other factors considered below, suggests that the continued rapid expansion of industry is uncertain unless the recent attempts to integrate the Latin American economies and plan industrial investment are successful.

Moreover, even if rapid industrialization continues it would be unrealistic to assume that it would solve the acute problem of unemployment which besets the region. As the data in the table below indicate, employment in manufacturing has hardly been able to keep up with the increase in the labour force—in spite of 35 years of quite rapid industrialization. It seems that during the early phases of growth the expansion of manufacturing activities tends to destroy the indigenous handicraft industries. These industries—producing clothing, footwear, simple utensils and household goods—are very labour intensive and their substitution by mass production techniques releases a lot of labour. In a later phase, in the somewhat more industrialized countries, 'modernization' of the manufacturing sector may occur through a process of substitution of highly mechanized methods of production (appropriate to the wealthy nations) for older and relatively less capital intensive techniques. In addition, union pressure and minimum wage legislation may raise the wage rate above the opportunity cost of labour and induce investors to adopt excessively mechanized techniques of production. This process clearly has occurred in Chile and Colombia.

[6] According to surveys sponsored by the UN Economic Commission for Latin America, nearly 30 per cent of the 'empresarios' in Argentina were born abroad, 41 per cent in Colombia were born abroad and 48 per cent in Chile were born abroad. (See ECLA, *El Empresario Industrial en America Latina*, 1963; E/CN. 12/642 and Additions 1, 3 and 4.)

[7] Indeed, selective emigration is a problem even in the wealthier Spanish American countries. During the period 1951–1961 over 8,500 professionals, technicians, high level administrators and skilled workers emigrated from Argentina to the United States. (See M. A. Horowitz, *La Emigracion de Profesionales y Technicos Argentinos*.)

Table I:3

EMPLOYMENT IN MANUFACTURING INDUSTRY AS PER CENT OF TOTAL EMPLOYMENT

	1925	*1960*
Argentina	20	21
Chile	21	17
Colombia	17	15
Peru	18	15
Venezuela	10	12

Source: F. H. Cardoso and J. L. Reyna, '*Industrializacion, Estructura Occupacional y Estratificacion Social en America Latina*', mimeographed, August 23, 1966, Santiago, Chile, p. 13.

The final source of growth of national output has been in the rapid expansion of the 'services' sector. This sector, however, is largely a sponge which absorbs the excess population of the rural areas[8] and its 'growth' may represent virtually no increase in economic welfare. That is, expansion of this sector may be a mere reflection of (i) the rapid rate of growth of the population, (ii) the increasing inability of the rural sector to provide employment opportunities, (iii) the pronounced migration from rural areas to urban slums[9] and (iv) the slow growth of employment opportunities in industry. Even if one would not want to follow Marxist national accounting techniques, and consider the services sector wholly unproductive, many of the apparent increases in the services sector should be ignored completely in calculating the rate of growth of national output. If this were done, Spanish American growth rates would be considerably lower than those reported in international statistical sources.

In Colombia, for example, the rate of growth of GDP appears to have remained unchanged since 1950. As the data in the table demonstrate, however, the aggregate growth rates obscure two offsetting tendencies: a sharp decline between 1960 and 1965 in the rate of growth of production of *goods* and an acceleration in the growth of services.

[8] In the decade of the 1950s considerably more than half the new additions to the labour force were employed in services in Argentine, Chile, Uruguay and Venezuela. In Colombia and Peru it was 49 per cent and 37 per cent respectively.

[9] See Table I:1 for data on the proportion of the population living in urban areas.

Table I:4

COLOMBIA: CUMULATIVE RATE OF GROWTH OF GDP
(average annual per cent increase)

	1950–1960	*1960–1965*
Agriculture and other primary activities	3·4	2·9
Manufacturing and construction	6·5	5·0
Services	5·0	5·6
GDP	4·5	4·4

Source: Banco de la Republica, *Cuentas Nacionales*, Bogota, Colombia.

The rate of growth of agricultural output in Colombia declined from 3·1 per cent to 2·6 per cent a year, i.e. substantially below the rate of increase of population. Growth of manufacturing activities declined from 6·5 per cent to 5·6 per cent a year. Services, on the other hand, grew much more rapidly than in the past. The growth of trade accelerated from 4·5 to 4·8 per cent; financial and real estate transactions accelerated from 6·5 to 8·8 per cent per annum; 'other services' expanded from 4·5 to 6·6 per cent per annum. Given the decline in the rate of growth of output of goods and the stagnant *per capita* income, it is difficult to believe that the apparent acceleration in the provision of services represents a real increase in output and economic welfare. The expansion of services is much more likely to be a reflection of the acute and growing problem of un- and under-employment.

This raises the question of valuing the output of services. One suspects that the recorded increase in the output of services is more apparent than real. In few service activities is there an independent measure of production; that is, output of services is estimated by valuing *inputs*. In Argentina the growth of services is determined from the change in the volume of inputs; in several other countries it is determined by multiplying the increase in employment by an estimate of the average wage; in some countries, e.g. Ecuador, the legal minimum wage, and not the lower actual wage is used.[10] In other words, because of the conventions used in national accounting, the index of services output might better be interpreted as an index of employment in services. Thus if tasks in agriculture formerly done by three men are now done by two, and in the services sector tasks done by four men are now done by five, the statisticians will claim

[10] See United Nations, *National Accounting Practices in Sixty Countries*, Series F, No. 11 (64. XVII. 9).

that productivity has risen in agriculture and output has risen in services. In fact, total output may have remained unchanged; all that may have happened is that the location of the underemployed labour may have shifted from rural to urban areas.

In a recent publication[11] the Economic Commission for Latin America has observed that productivity is declining in several important sub-sectors of the economy, *viz.* government, construction, trade and commerce, 'other' services and 'miscellaneous' services. Over the period 1950–1962 product per worker declined 0·3 per cent per annum in construction and 0·2 per cent per annum in services throughout Latin America. Within the services sector, productivity in government, miscellaneous services and unspecified activities is estimated to have declined 1·0, 0·7 and 0·8 per cent per annum respectively. This is what our hypothesis would predict. Moreover, even in sectors in which productivity has increased, misplaced aggregation may hide important differences in the behaviour of the various groups comprising the sector; in fragmented economies one must not assume that sectors are composed of homogeneous units. It is highly likely that substantial increases in productivity have occurred in the large firms of the services sector while productivity has declined in the small firms. The distribution of income within the services sector probably is becoming more unequal; output per man year (and hence real income) of the majority probably is falling. To them growth is illusory.

The slow rate of growth of the agricultural sector has been the principal factor restraining development. It can easily be demonstrated with simple arithmetic that if the agricultural sector is both relatively large and slow growing it is extremely difficult for *per capita* GNP to persistently expand at a rapid rate. Assume, for example, that at a given moment in time agriculture accounts for 50 per cent of GNP and is growing at a rate of 2 per cent per annum; services represent 35 per cent of GNP and are growing at 3 per cent per annum. Under these circumstances a 5 per cent over-all growth rate can only be achieved if the industrial sector grows by 20 per cent per annum. Quite clearly, rapid economic progress cannot be achieved in Spanish America unless the agricultural sector begins to grow more rapidly; it is precisely the poor performance of this sector which is responsible for the low over-all rate of growth.

Of course, there has been some increase in productivity in agriculture. Within the sector, however, the rise in output per man-hour has not been uniform: productivity on the minifundia probably has

[11] *Economic Survey of Latin America 1964*, 1966, pp. 43–6.

fallen while productivity on the latifundia probably has increased. To the extent that this has occurred, growth has been illusory to the majority of rural inhabitants.

There is considerable evidence that the distribution of income in agriculture is becoming more unequal. In Chile the share of wages has declined and the distribution of irrigated land has become more unequal.[12] In Argentina the agricultural wage share fell by nearly 28 per cent between 1950–54 and 1959–61 and real earnings in agriculture (and industry) declined after 1955.[13] Even in Peru, where output clearly has increased more rapidly than the population, large groups in the society have not shared in the benefits of growth: between 1954 and 1959 *per capita* income on the Peruvian coast rose 5 per cent, whereas it fell by 7 per cent in the Sierra.[14] The reasons for this are examined in detail below.

Regional Dynamics

Most economists implicitly assume that the growth and development of the rich nations and regions are independent of the stagnation and underdevelopment of the poor nations and regions. Our growth models and development theories, almost without exception, assume that at one point all the nations were more or less underdeveloped. Some of the nations 'took-off' and developed, leaving the others behind. The problem is to examine the rich nations, try to understand what enabled them to grow, and then adapt the conclusions to the 'backward' countries so they can 'catch up'.[15]

This, as was argued at length in the Introduction, is largely a false picture. The vast majority of today's non-communist nations participate in a single, integrated world economic system—and have done so for centuries. This system may be called industrial and mer-

[12] Inter-American Committee for Agricultural Development, *Tenencia de la Tierra y Desarrollo Socio-Economico del Sector Agricola en Chile*, 1966, pp. 30, 10.

[13] This information was kindly provided to me by Mr Victor Tokman of St Antony's College, Oxford.

[14] See R. Thorp, 'Inflation and Orthodox Economic Policy in Peru', *Bulletin*, of the Oxford University Institute of Economics and Statistics, August, 1967.

[15] Harvey Leibenstein's writings are almost a caricature of this approach. See his *Economic Backwardness and Economic Growth*, p. 4. In general, speculations as to the cause of a nation's economic growth have centred on such items as its endowment of natural resources, the 'savings coefficient', parental attitudes towards child-care and their effect on the supply of entrepreneurship, 'technical change', education, etc. All of these hypotheses include the assumption that growth of one nation is largely independent of its relations with another.

cantile capitalism. A major feature of this system is that the relationships between its various components (nations in the world economy or regions in a national economy) are unequal, and thus the benefits of participating in the system are unequally and inequitably distributed.[16]

The distribution of the fruits of intercourse may be such that in the prejudiced area (a) the level of consumption falls and the region is absolutely worse off, (b) consumption increases once-for-all, but savings and hence the rate of growth of output decline, or (c) only the relative rate of growth declines.

The important point, as Myrdal has stressed,[17] is that in a *laissez-faire* environment in which unguided market forces determine resource allocation cumulative movements in income inequalities are likely to be set up. This will occur not only between nations,[18] but also within a customs union[19] as well as between regions of a single nation. The growing inequality in the international distribution of income has received considerable comment. Studies of regional disparities in income within the under-developed countries, in contrast, have been relatively less frequent.[20] Of course, many observers have noted that the Santiago area is growing much faster than the agricultural regions of Chile, for example, but few have become interested in studying the general phenomenon of regional interdependence. There has been a proliferation of regional plans from the Cauca Valley in Colombia to the Mesopotamia in Argentina, but their approach essentially has been to consider the region as an *independent* unit and to determine, say, what institutional changes and additional investments would be necessary to increase the region's growth rate by a predetermined amount.

Yet it is quite likely that broader studies of inter-regional relations would indicate that the relative poverty of some areas is due directly to the type of associations experienced with other areas. Furthermore, the study of inter-regional trade—where market forces generally are unhampered by controls or obstacles—could provide useful evidence as to how our (essentially capitalistic) economic system works.

[16] See T. Balogh, *Unequal Partners*, Vol. I.

[17] See G. Myrdal, *Economic Theory and Under-developed Regions*.

[18] See K. Griffin and R. Ffrench-Davis, *Comercio Internacional y Politicas de Desarrollo Economico*, 1967, Ch. II.

[19] See Ch. VI below.

[20] See, however, P. R. Odell, 'Economic Integration and Spatial Patterns of Economic Development in Latin America', *Journal of Common Market Studies*, March 1968.

The Case of the Peruvian Sierra

A strong hint that regional interdependence may be an important factor in explaining the poverty of some areas is provided by the case of Peru. It is quite clear that internal migration, trade flows, and capital movements may have had *absolute negative effects* on the level of *per capita* consumption and the rate of growth of the poorer region.

Internal migration has resulted in a transfer to the cities of the most ambitious and skilled rural workers. Precisely the most valuable human resources of the countryside are lost to the urban areas, where they may spend up to ten years searching for work in the factories. The government's policy of linking the coast and the Sierra with highways, by facilitating the exodus has only aggravated the problem. As the authorities themselves have recently recognized, '*cuanto mas carreteras se construyeron, mas multitudes de serranos bajaron a la costa, principalmente a la cuidad de Lima*'.[21]

Even more important than the transfer of labour to the coast has been the transfer of capital. The inter-regional trade figures are eloquent in this respect.

Table I:5

INTER-REGIONAL TRADE BETWEEN THE PERUVIAN SIERRA AND THE COAST, 1959
(millions of soles)

	Imports of the Sierra from the coast	Exports of the Sierra to the coast
Agricultural products	174	3,002
Minerals	198	459
Industrial products	671	473
Services	233	0
Commerce	476	99
Finance	2	30
Other	89	131
Total	1,843	4,194
Sierra's export surplus	2,351	
	4,194	4,194

Source: Banco Central de Reserva del Peru, cited by Barandiarán, L., '*Estrategia de desarrollo económico en la Sierra Peruana*', 1964 (unpublished).

[21] Comision Ejecutiva Interministerial de Cooperacion Popular, *El Pueblo lo Hizo*, año 1, No. 1, September–December 1964, p. 27.

Per capita income on the coast is approximately $520, or roughly 6½ times higher than that of the Sierra. The above table indicates that the Sierra had a surplus of over 50 per cent in its trade with the coast, i.e., that its output exceeded its expenditure by a considerable margin. This means that it is very probable that its *level of consumption was lower than it would have been had there been no trade.*[22] Orthodox theory would lead one to think that the export surplus would be compensated on capital account through the accumulation of deposits (and other assets) in the region's banks. These deposits would constitute savings for the region and accelerate its rate of growth.

The orthodox presumption, however, is the reverse of the truth. The Sierra's agricultural exporters are latifundistas who live mostly in Lima and deposit their receipts in the capital's banks. Thus the savings generated in the Sierra are transferred to the coast and as a result, assuming investment occurs where savings are held, the *rate of growth of the Sierra probably is lower than it would have been in the absence of trade.*[23] Indeed, as we have seen, the level of *per capita* income declined in the Sierra.

Peru's banking system, its land tenure arrangements, and its social institutions are so organized that savings are siphoned from the poorer region to the rich one. This exploitation of the poor by the rich through the mechanism of inter-regional trade is analogous to the relationship which existed (and in some cases still exists) between the metropole and the colonial territories. In fact, Latin America has been described as a region dominated by '*colonialisme intérieur*'.[24]

In the case of Peru the consequences of the savings transfer have been disastrous for the rate of growth of the Sierra, as the export surplus represents about 16·8 per cent of its gross regional product. This transfer was sufficient to reduce the annual rate of growth of

[22] One could argue that in making welfare comparisons consumer surplus is the relevant concept, not the market value of the transactions. That is, if the intra-marginal valuation of manufactured goods in the Sierra is extraordinarily high, it is conceivable—but in practice hard to imagine—that the Sierra could gain by exchanging 4,194 million soles of mostly food for 1,843 million soles of services, commerce, and industrial products. At the very least, one must agree that the distribution of the gains from trade has been most unequal.

[23] This is not, of course, an argument against trade; it is an argument in favour of changing the institutional relations which affect the distribution of the gains from trade.

[24] See R. Dumont, *Terres Vivantes*, Ch. 1. The phrase was originally coined to describe Colombia, but it is equally applicable to the other Spanish American countries.

the Sierra by 4 or 5 precentage points. From the point of view of the coast the capital transfer represents slightly less than 4 per cent of its gross regional product, although the contribution to the rate of capital formation on the coast is considerably greater, *viz.* 15 per cent. Hence it appears that inter-regional relations have impeded the development of the Sierra, although it is possible these relations may not have been of crucial significance to the development of the coast.[25]

Evidently the process of development and underdevelopment are related; they are not independent of one another. It is the manner of integration and the resulting unequal distribution of benefits which in large part may explain the poverty of some nations or regions and the development of others. The predominance of 'backwash effects' in some regions probably is closely associated with the degree of concentration of economic and political power. Some of these issues will be explored in the remaining parts of this chapter.

2 SOCIAL CONFLICT AND ECONOMIC POWER

In the poorer countries of the world one finds that class and group conflicts are much sharper than in the industrialized nations. A partial listing of such conflicts in Spanish America might include:

agricultural exporters *v.* industrialists demanding protection
landless labourers *v.* latifundistas
minifundistas *v.* monopsonistic middlemen
unemployed urban workers or *Lumpen Proletariat v.* high wage union labour.

With societies characterized by these devisive and enduring coalitions—as opposed to shifting coalitions—it is difficult to achieve a national consensus.[26] Development policy in such countries cannot

[25] That is, the Marxist thesis which affirms that the development of the metropole depends upon the exploitation of the periphery cannot be tested with the limited information available. It would seem, however, that the time has come to relax Kuznets' 'basic working assumption—that we can study the characteristics of (modern economic) growth most effectively if it is not merged with . . . the growth of such parts of the world as have not yet succeeded in tapping the sources of modern economic growth'. (S. Kuznets, 'Notes on the Take-Off', in W. W. Rostow, editor, *The Economics of Take-Off into Sustained Growth*, p. 22.)

[26] The US has often been described as a 'nation of joiners' in which a large proportion of the population are members of several organizations whose aims duplicate and cut across one another. This has led to a society of shifting coalitions

be purely technocratic nor can planning be merely 'indicative'; development programmes must attempt the much more arduous task of coupling economic expansion with rapid and profound social and institutional transformation. The insistence on the importance of fundamental changes is not solely a question of political preference or of ideology; in many cases it is an urgent necessity if the nation is to avoid the extremes of violent civil discord or repressed violence under a military dictatorship. Violence already has become rather common in at least four Spanish American republics, namely, Bolivia, Colombia, Venezuela and Peru, and it could easily spread to the others. Greater violence would be a tragedy: it would probably delay indefinitely the introduction of reforms and those who attempt it are almost certain to be destroyed. Force is most unlikely to achieve its objectives, but this does not mean it will not be tried by desperate men.

Unfortunately, many sociologists (as well as economists) tend to view society as a self-equilibrating model in which chronic disequilibria are a special case. This is particularly true of the Talcott Parsons 'structural-functional' school which predominates in the United States and many UN agencies. In general, the 'equilibrium model' of society is not very useful, and it should be replaced by a conception of society as a 'tension-management system' in which order itself is problematical rather than assumed.[27] In such a context the task of planning and development policy cannot be limited to increasing *per capita* consumption, or even providing for its equitable distribution. It must also foster democratic participation in the decision-making process so that the various groups in the nation associate their own interests with the stability of society as a whole.

Social Dualism: a False Hypothesis

One must be careful, however, not to assume that social tension and a poorly articulated economy imply the presence of a dual society.

in which class and interest boundaries are softened and blurred. On the other hand, in the underdeveloped countries—and to a lesser extent, in Europe—one does not 'join' a class or group, one is 'born into' it. This strengthens class consciousness and accentuates divisive tendencies in society. A similar pattern developed in the United States with respect to the Negro minority and led to the 'Negro Revolution' we are now witnessing.

[27] See W. E. Moore, 'Predicting Discontinuities in Social Change', *American Sociological Review*, June 1964, p. 337. For a criticism of the applicability of modern economic theory to the problems of under-development see D. Seers 'The Limitations of the Special Case', *Bulletin* of the Oxford University Institute of Economics and Statistics, May 1963.

It is frequently affirmed that colonialism consisted of the superimposition of a European society upon an indigenous social structure, and that the latter *continued its existence undisturbed and unchanged.* The problem of development then is viewed as 'integrating' the unchanged, backward, and 'traditional' sector into the modern economy. Thus, for example, a group of UN economists, sociologists, and political scientists asserts that 'The *social structure* of Latin America has in the past been characterized by a serious lack of integration.'[28]

Yet the effect of colonialism was not to isolate but to *destroy* the indigenous social structure and to re-integrate the original population into a capitalist-colonialist[29] system which was and is highly unfavourable to their interests. This system has persisted in Spanish America even after 150 years of independence. Referring to Peru, one writer states:

> 'The new landowners introduced a highly commercialized economic system based on the use of money and on competition in the international market, where no market existed before, and they consolidated their power through the hacienda, or plantation estate. By these changes the original population was reduced to a state of social and economic disrespect which persists to the present day.'[30]

The degree of 'disrespect' to which the original population was reduced is obvious to every tourist who has been to Cuzco and Machu Picchu. 'The economic system of the Indians was wiped out by the Spaniards, together with the culture which it supported, . . . but it was not succeeded by an economy of greater productiveness. Indeed as is evidenced by the numerous terraced hillsides, intensively cul-

[28] 'Report of the Expert Working Group on Social Aspects of Economic Development in Latin America', *Economic Bulletin for Latin America,* Vol. IV, No. 1, March 1961, p. 56. Emphasis in the original.

[29] The common claim that Latin America is 'feudal' can be highly misleading. From the beginning of the Spanish conquest the region was incorporated into a world-wide mercantilist system. This was true not only of Mexico and Peru (which exported precious metals) but also of Portuguese Brazil. Celso Furtado indicates that from the sixteenth century onwards the Brazilian economy was essentially capitalist, being based on specialization and the division of labour (sugar), reliance on foreign markets, and investments in slaves. See C. Furtado, *Formacion economica de Brazil,* Fondo de Cultura Economica, Mexico, 1962, pp. 58–9. Translated from the Portuguese (*Formacao economica do Brazil*) by Demetrio Aguilera Malta. For a general discussion see S. Bagu, *Economia de la Sociedad Colonial,* especially Ch. V.

[30] A. R. Holmberg, 'Changing Community Attitudes and Values in Peru: a Case Study in Guided Change', *Social Change in Latin America Today,* p. 55.

tivated before the Conquest and now completely abandoned, productivity declined in many areas.'[31]

Hence it appears that any programme designed to improve conditions for the underprivileged mass of the population must concentrate not so much on integrating these people into society as on *changing their relationships with the rest of society*.[32] One is not trying to merge two different cultures but to alter the terms in which the separate groups of a single national society engage in social, political, and economic intercourse.

Counter-force v. *Community Development*

It has become quite evident that 'planning from above' is impossible and if progress is to occur, mass participation in the development process must be encouraged. Such programmes, at least in theory, have become increasingly popular. In practice, most of the efforts to mobilize the masses have centred on the rural community or urban residential neighbourhood, and the object of the programmes has been to assimilate or integrate the community into the national society. A typical example is the Peruvian programme of *Cooperación Popular* which speaks of '*reincorporando grandes grupos que han sido olvidados, al incorporar la communidad campesina a la vida de la nación*'. The programme '*se apoya en la solidaridad del grupo humano* . . .' and a public organization has been created which is in charge of the '*ejecución de obras públicas que se realicen por la acción voluntaria del pueblo*. . . .'

Thus the community development approach to democratic organization assumes not only that non-integrated social groups co-exist, but furthermore, that there is a 'harmony of interests'[33] both *within* the community and *between* the community and other regions of the nation. The analysis in the previous sections would seem to indicate that, on the contrary, (a) the society is already integrated—although to the disadvantage of many of its members, (b) the various groups

[31] R. E. Christ, 'The Indian in Andean America, I', *The American Journal of Economics and Sociology*, Vol. 23, No. 2, April 1964. In spite of the fact that the Inca economy and culture were 'wiped out', Christ persists in claiming that the Spaniards 'superimposed' the hacienda system upon the Indians.

[32] It is not the quantitative aspects of the relationships which interest us here (extent of autoconsumption, value or volume of sales or purchases) but their nature or quality.

[33] The assumption of 'harmony of interests' is analogous to the doctrine of *laissez-faire* in which the welfare of society is believed to be maximized if each individual pursues his private interest. This, as T. Balogh once remarked, is a doctrine designed to make those who *are* comfortable, *feel* comfortable.

of a community may have sharply differing interests, and (c) there may be serious conflicts between the interests of one region and those of another. In particular, it would be very ill-advised to assume that the interests of the latifundistas, minifundistas, landless labourers (e.g. the inquilinos in Chile), seasonal workers (afuerinos), and merchant money-lenders of a rural 'community' are identical, or even compatible. In many cases the rural 'community' in Spanish America is synonymous with the latifundia; agricultural villages are not as predominant as in Asia.

The experience with community or village development programmes in Asia—where the UN has strongly advocated them, as well as the author's observations in Algeria, show that such projects have not been successful in raising agricultural production, stimulating village industry, extracting savings from the rural sector, or using surplus labour efficiently.[34] Evidently, future policy in this area will have to be based on a different theoretical framework. It would seem that rather than viewing the problem of popular participation in development as one of merging non-integrated societies so that their common interests become apparent, the opposite approach, namely, viewing society as composed of various groups with opposing interests, would be more enlightening.

Such an approach would lead one to search for the *conflicts* underlying the relations between groups, the coalitions and *bargaining* to which they give rise, and the *monopoly* elements pertaining to each group which determines the outcome of the process.[35]

If one looks at Spanish American society one finds that each group tries to carve out a monopoly position for itself. Capital markets are tightly controlled by a small minority[36] to which most members of society have scant access, and on the occasions when they *are* granted access they are forced to pay exorbitant rates of interest. It is not uncommon for the monthly rate of interest paid by small borrowers to approach 10 per cent, and in Ecuador cases have been found of the Catholic Church charging 200 per cent per annum on

[34] See UN, *Community Development and Economic Development*, Part I, *A Study of the Contribution of Rural Community Development Programmes to National Economic Development in Asia and the Far East.*

[35] See K. H. Silvert, *The Conflict Society: Reaction and Revolution in Latin America.*

[36] The Chilean case has been well documented. See Instituto de Economia, *Formacion de capital en las empresas industriales;* R. Lagos, *La Concentracion del poder economico.*

loans to peasants.[37] In Chile, interest rates from rural money-lenders are about 27–128 per cent a year for loans in cash and about 60–90 per cent a year for loans in kind.[38]

The Marxists have assumed the financial and industrial monopolists would be firmly opposed—and eventually defeated—by the solidarity of the working class. One finds, however, that it is the more privileged members of the working class—civil servants, professionals (including economists), and skilled industrial labourers—who are best organized. Less than 10 per cent of all the workers in Chile, for example, are organized, and the percentage in the less industrialized Spanish American economies is much lower. The organized groups are anxious to protect their position and restrict entry to their job categories. Their interests differ sharply from those of the unorganized and the under-privileged members of society; there is no class solidarity.

The Chilean copper miners, for example, receive incomes roughly five times higher than those of other industrial workers in the country. For this reason the socialist labour unions in the mines appear to be unenthusiastic about nationalizing the foreign-owned companies and risking a relative decline in their favoured position. Furthermore, even within the copper industry there is little worker solidarity. Workers in the former Braden Co.'s mine earn less than Anaconda's workers. When the latter struck in 1963 the former went out in sympathy, forcing the government to apply more pressure for a settlement at Anaconda. When Braden went on strike six months later, Anaconda's union 'discussed' going out in sympathy, but never did, and as a result the wage disparity within the industry increased.

Monopoly of Land and Exploitation of Labour

The rural workers are by far the most numerous and miserable of the under-privileged groups and their relative position has continued to deteriorate. The United Nations Economic Commission for Latin America states, '*La estratificación rural mostraba una tendencia marcada a la polarización de la población en dos grupos desiguales hacia los extremos de la escala*'.[39] Given (a) the high population

[37] Rafael Baraona, *Rasgos fundamentales de los sistemas de tenencia de la tierra en el Ecuador*, A Report prepared for the Inter-American Committee for Agricultural Development, preliminary version, 1964, pp. 133, 186–7, 267.

[38] C. T. Nisbet, 'Interest Rates and Imperfect Competition in the Informal Credit Market of Rural Chile', *Economic Development and Cultural Change*, October 1967.

[39] UN, ECLA, *El desarrollo social de America Latina en la post-guerra*, E/CN. 12/660, p. 29.

growth rate, (b) the low rate of growth of output and the even smaller increase in food production,[40] and (c) the worsening distribution of income in rural zones, it is quite probable that consumption levels in the lower strata of society are falling.[41]

This deterioration of the economic position of the most vulnerable members of society is due in large part to the lack of bargaining power of these groups. These people are unorganized and dispose of little or no capital or land. Furthermore, they face a monopsonistic labour market.

As Frank has stated, 'the key source of the great inequality of bargaining power in the labour market is unquestionably the extensive ownership of land on the part of the few and the absence of or severe limitation in the ownership of land on the part of the many'.[42] The extreme to which land has been monopolized in Spanish America is presented in the table below. Unlike most measures—which compare the size distribution of land among landowners—the data below are estimates of the size distribution of land among the active agricultural workers. This should give us a more

[40] *Per capita* food production is lower today than it was a decade ago in Argentina, Chile, Colombia, Peru and Uruguay. (See FAO, *The State of Food and Agriculture*, 1967). The record for agricultural production as a whole is equally as bad and, as the table below indicates, the problem has existed for a long time.

YEARLY PERCENTAGE RATE OF INCREASE IN AGRICULTURAL PRODUCTION AND POPULATION, 1945–47 TO 1958–60

	Agricultural Production	Population
Argentina	1·0	2·1
Bolivia	1·3	2·0
Chile	1·8	2·2
Ecuador	7·2	3·0
Paraguay	1·5	2·4
Peru	2·9	2·3
Uruguay	1·4	1·6
Venezuela	4·6	3·7

Source: ECLA, *Agriculture in Latin America: Problems and Prospects*, April 6, 1963 (E/CN. 12/686), mimeo., p. 5.

[41] Additional support for this hypothesis is provided by a UN study which indicated that housing conditions in many countries are deteriorating, at the same time that expenditure on luxury apartments in the capitals continues unabated. See ECLA, *Provisional Report of the Latin American Seminar on Housing Statistics and Programmes* (E./C.N., 12/647, February 1963).

[42] A. G. Frank, '*La participación popular en lo relativo a algunos objectivos económicos rurales*', p. 32, mimeo., no date. Also see his *Capitalism and Underdevelopment in Latin America*, Ch. II.

relevant indication of the extent to which the agricultural producers are able to dominate the labour market, and it is found, for example, that in Venezuela roughly 1 per cent of those active in agriculture own nearly 75 per cent of the land.

Table I:6

DISTRIBUTION OF LAND

	Per cent of active agricultural workers	Per cent of land owned*	Per cent of agricultural families without land or owners of less than 5 ha.
Argentina	1·780	74·9	64·6
Chile	0·503	73·2	73·4
Colombia	0·157	26·7	n.a.
Ecuador	0·110	37·4	75·4
Paraguay	1·020	86·7	n.a.
Peru	0·098	76·2	n.a.
Uruguay	1·250	56·2	67·9
Venezuela	0·960	74·5	90·6

Source: S. Barraclough and E. Flores, '*Estructura agraria de America Latina*', *Informe del curso de capacitacion de professionales en reforma agraria*, Vol. I, pp. 279–80 (Instituto de Economia, Universidad de Chile, 1963). The data refer to various years between 1950 and 1960. Data for Paraguay obtained from D. Kanel and C. Fletschner, '*La Distribucion de Tierras en el Paraguay*', mimeo., LTC No. 24, Land Tenure Center, University of Wisconsin, August 1966. The figures for Paraguay, unlike the other countries, do not include landless workers, but they do include squatters.

* Refers to holdings over 1,000 ha.

Inequality in the distribution of land (and water) is, of course, reinforced by an unequal distribution of educational opportunities. This monopolization of the complementary elements of production has led to the creation of a vast pool of unskilled, mostly unemployed, and almost unemployable, agricultural labour. The supply of manpower available to the latifundistas (and urban industrialists) is greater, and its price consequently lower, then would occur if the monopoly elements were not present, i.e. if holdings were small and self-employment greater. In addition, concentration of resources tends to increase the under-utilization of latifundia land and thereby lower the demand for labour.[43]

[43] Even in the Argentine Pampa, the most productive and efficiently cultivated region in Spanish America, only 48·7 of the agricultural land is cultivated. The remainder is in fallow land, natural pastures (41 per cent), or woods and brush. On the large 'estancias' an even smaller percentage of the land is cultivated. See

From the point of view of the landowner this system is highly profitable. Case studies in the Ecuadorian Sierra indicate that the average hacienda earns 33 per cent gross profit (excluding amortization) on its sales receipts.[44] The reason why the returns are so high is that most factor inputs are provided gratis or nearly so. In the majority of cases labour is provided by *huasipungueros* who work four days a week for the hacienda and, on those days, receive a nominal wage.[45] In addition, the *huasipungueros* are allotted a small plot of land to cultivate. These workers frequently must provide their own houses and hand tools. Additional labour is provided by *yanaperos* who, in return for the right to use the hacienda's roads, collect water for domestic purposes, and gather firewood, agree to provide several days' work gratis. The *yanapo* or *pago de los pasos* also includes the use by the hacienda of the flocks of the indigenous people to fertilize its fields. Thus even fertilizer is free!

A summary of two typical cases from the Ecuadorian Sierra is presented in the table below.

The table and the two accompanying notes clearly indicate how

Table I:7

TWO LATIFUNDIA IN ECUADOR

(1) case	CV(3)	GB(9)
(2) money income	583,184 sucres	560,924 sucres
(3) money expenditure	196,028 sucres	336,935 sucres
(4) of which, wages and salaries	136,151 sucres	89,163 sucres
(5) profits (excluding amortization)	387,165 sucres	223,990 sucres
(6) profits as a percentage of income	66·5 per cent	39·9 per cent
(7) workforce	101* persons	114 persons
(8) size of hacienda	690 ha.	12,000 ha.†

Source: Baraona, R., *op. cit.*, pp. 99–110, 156–73.

* Most of the work is provided gratis by 45 'yanaperos'.
† 99·5 per cent of the land surface is not cultivated

Arthur L. Domike, *Land Tenure and Agricultural Development in Argentina*, A Report prepared for the Interamerican Committee for Agricultural Development, preliminary version, 1965, p. 25.

[44] R. Baraona, *op. cit.*, pp 76, 89, 110, 146, 173 and 196.

[45] The typical wage is 3 sucres per workday, but this is paid only sporadically. (The sucre/dollar exchange rate was 18·9/1 at the time of the study.) Wages are deducted and fines imposed for any infraction of the hacienda's rules, e.g. if the huasipunguero's animals graze on the hacienda's lands. Consequently, money seldom changes hands between the patrón and his huasipungueros.

profitable it can be to monopolize the land and exploit labour. It should not be difficult to understand why agricultural techniques of production have not changed radically from one century to another.[46] It is hardly in the interests of a profit maximizing latifundista to do so.

It appears that the small tenant frequently is little better off than the agricultural labourer. The tenancy arrangements to which he must submit are onerous (small plots, obligations to contribute labour to the patrón, crop sharing) and provide little opportunity for achieving a decent standard of living. The possibilities of escaping from this system are slight since the tenant neither owns capital nor can accumulate it. Tenants often are compelled to sell their crop to the landlord at the harvest period. The landlord is then able to reap an additional profit when he resells the crop at a higher price at a later date. These monopsonistic purchasing arrangements may be disguised as capital loans: the tenant is kept in heavy debt and 'repayments' are made through earmarking his future production (valued at harvest time prices) as security for new loans. In some countries the lender is often the small town merchant rather than the landowner. The merchant supplies the seed and some cash. In return he receives 50 per cent of the harvest and the right to sell all of it, thereby reaping a profit on the other 50 per cent as well. In other cases the extortion is less subtle. Baraona tells us that '*En el mercado local, es costumbre, sobre todo en los días de feria* (*domingos*) *que los pequeños comerciantes del pueblo acudan a los caminos de la comarca a comprar, incluso violentamente, los productos que entren en el mercado, consiguiendo así, y por su calidad de hombre blanco, majores precios*.'[47]

Thus the weak are exploited by the strong, both in the labour market and in the commodity market. The way to redress this exploitation is to increase the bargaining strength of the former relative to the latter, to establish a counterforce by mobilizing political support for the less favoured members of society and greatly increasing their command over resources. One will get nowhere assuming there is an inherent convergence of interests among the various groups comprising a rural community.

[46] See however, T. W. Schultz, *Transforming Traditional Agriculture*, p. 174 (also p. 169). 'Why many of the farmers who own and are responsible for the operation of very large farms, especially in some parts of South America, do not engage successfully in this search for modern agricultural factors is a puzzle.' Puzzle indeed!

[47] R. Baraona, *op. cit.*, p. 179.

3 ECONOMIC CONSEQUENCES OF THE LATIFUNDIA SYSTEM

Before considering policies which might be introduced to correct the social system we have described, the economic consequences of the latifundia system should be examined in greater detail. Perhaps the best way to do this is to discuss how it functions in a particular country, say, Colombia.

For at least the last seven years agricultural production as a whole in Colombia has increased less than the population, with the result that the *per capita* availability of agricultural products has declined by about two per cent. Livestock production, however, has grown considerably faster than the population, while crop production has lagged seriously behind.

Not all crops, of course, have grown at the same rate, and to fully understand what has been happening in Colombia one must disaggregate the data. In general, with the exception of coffee (which has increased very slowly) and cotton (which expanded very rapidly but has recently encountered technical difficulties), the production of export crops has advanced fairly smoothly, e.g. bananas and sugar cane. Production of food for domestic consumption, on the other hand, has, in many instances, become virtually stagnant. In other words, the great cause of the poor performance of agriculture as a whole has been the slow increase in supply of basic foodstuffs.

The food supply problem, in turn, is intimately related to the way

Table I:8

COLOMBIA: VOLUME OF AGRICULTURAL PRODUCTION, 1958–1965
(per cent increase over the period)

Total agricultural output	20·5
Livestock	27·4
Crop production	16·4
Domestically consumed Food Crops	
Rice	39·1
Beans	−40·3
Maize	5·8
Potatoes	8·9
Plantain	22·5
Wheat	−28·6
Manioc	20·0
Population	23·0

Source: Banco de la República, *Cuentas Nacionales*, Bogotá, Colombia.

the land is used and the closely associated land tenure conditions. According to the agricultural census of 1960, over half of the agricultural land in the Andean and Caribbean regions of Colombia is in fallow or natural pastures, while only 15·7 per cent is cultivated. In other words, nearly four times as much land is devoted to extensive grazing of cattle as is devoted to crops. Moreover, the cattle frequently occupy the relatively flat lands in mountainous areas and, in company with the large-scale commercial agricultural enterprises producing for export, the broad river valleys; food production for domestic consumption is largely confined to the steeper slopes of the hills and mountains.

Agricultural land is highly concentrated in a few hands. Data in the CIDA report referring to 1960 indicate that in the Andean region 5·3 per cent of the agricultural exploitations account for 67·9 per cent of the land, while at the other extreme, farms of under 5 ha., which represent 63·5 per cent of the exploitations, account for only 1·7 per cent of the land.[48] The distribution of land in the less densely populated Caribbean region is equally as skewed.

Even these figures, however, greatly understate the extent to which the land is controlled by a relatively small number of individuals. First, the above data refer to the distribution of farm size, not the distribution of farm ownership. Many persons may own more than one farm or exploitation. For example, in the Andean and Caribbean regions there are only 68 farms larger than 5,000 ha. and these account for 660,000 ha. in all. On the other hand, there are 202 landowners who possess more than 5,000 ha., and these account for 5·8 million ha. in all.[49]

Secondly, the data on the distribution of land by farm size fail to take into account the large and rapidly growing number of landless agricultural workers. Some of these labourers become permanent hired hands on the larger exploitations while an increasing number are forced to become migrants, moving from one harvest to another. The authors of the CIDA report indicate that in 1960 about 8·6 per cent of the agricultural labour force was landless;[50] the proportion is substantially higher today and some observers estimate that 300,000 families, i.e. over 1·5 million people, are in this position. These people have a standard of living roughly comparable to a minifundista and for many purposes landless labourers and minifundistas can be considered members of a single group.

[48] CIDA, *Tenencia de la Tierra y Desarrollo Socio-Economico en el Sector Agricola en Colombia*, p. 72.

[49] *Ibid.*, p. 78.

[50] *Ibid.*, p. 135.

The influence of the tenure system upon the way land is used is clearly seen in the table below. The minifundistas tend to cultivate a very large proportion of their land and to leave little over a third in pastures and woods—in spite of the fact that their land frequently is located on very steep and eroded terrain. The latifundistas, in contrast, cultivate less than 6 per cent of their land and devote two-thirds of the remainder to livestock. In other words, most of the food produced in Colombia is provided by a large number of small farmers who between them own only a fraction of the agricultural land. Over 70 per cent of the agricultural surface is held in the form of latifundia, yet these farms account for only 28 per cent of the land under cultivation.

Table I:9

THE USE OF AGRICULTURAL LAND IN COLOMBIA, 1960
(per cent)

	Total Agricultural land	Cultivated surface	Fallow and natural pastures	Woods and mountains
Minifundia	100	62·5	32·0	5·5
Family Farms	100	29·1	49·4	21·5
Latifundia	100	5·7	65·8	28·5

Source: Based on data from CIDA, *op. cit.*, p. 139.

In social terms, the land tenure system has led to privilege and great inequalities in income and wealth. In economic terms, the tenure system has led to inefficiency and slow growth, to both underutilization of most of the land and underemployment of labour. The poor utilization of land and labour is further aggravated by the concentration of nearly all other resources in the hands of those least inclined to make productive use of them. Fifty per cent of the area under irrigation, largely financed by public investment, pertains to farms larger than 200 ha.[51] At the same time the data indicate that the farms of less than 5 ha. are 2·6 times more likely to irrigate than farms over 200 ha.[52] Similarly, 52 per cent of the agricultural credit goes to only 10 per cent of the borrowers.[53] In fact, as an analysis

[51] Ministerio de Agriculture, *Plan Cuatrienal Agropecuario 1967–1970 Para Ocho Productos de Consumo Popular*, Serie de Planeamiento No. 1, Feb. 1967.

[52] CIDA, *op. cit.*, p. 167.

[53] C. A. Fernandez, *La estructura del Crédito Agrícola en Colombia*, 1962, p. 41–2.

of the Caja Agraria will show, the credit system in Colombia can be understood only if one is familiar with social and political conditions in the country.

The Caja Agraria was established by the government in order to promote crop and livestock production and to make credit available to small farmers. It makes both short- and long-term loans for a variety of purposes, including harvest, finance, rural housing, and the purchase of agricultural implements. The rate of interest charged by the Caja varies, but on average it is 9·5 per cent. Considering that inflation in Colombia usually is about 10 per cent per year, this amounts—in real terms—to providing capital gratis.

A large proportion of this free capital is channelled to farmers who are by no means poor. The Caja is prepared to make general agricultural loans to any farmer, provided his investment capital does not exceed 1·5 million pesos. It will finance 'economic crops', e.g. cotton and sugar cane, for farmers with up to 3·0 million pesos in capital. Farmers with a maximum of up to 4·0 million pesos can obtain loans to mechanize their haciendas, and there is no capital limit for farmers who wish to finance irrigation schemes.[54] These limits are very generous indeed. For example, assuming a rate of return of only 5 per cent,[55] 1·5 million pesos would yield a family income of 75,000 pesos, i.e. roughly five times the national average.

Not only in theory is the Caja permitted to lend to the relatively prosperous; in practice it does so. This can be seen by comparing the average size of loan for crops typically produced by small farmers with the average loan for activities customarily undertaken by large farmers. Thus, for example, the average loan for cotton was 19,600 pesos, while the average coffee loan was only 2,800 pesos. Tobacco loans were little more than two thousand pesos on average, whereas loans for cattle were nearly three times as large.

The Caja is able to finance a large volume of credit, since 53 per cent of the institutionalized savings in the country are deposited in its savings department. The average deposit is very small, *viz*. 320 pesos, and these represent the liquid savings of the poor.[56] The Caja pays an interest rate of 4 per cent on its deposits; that is, in real terms it pays its creditors a rate of approximately minus 6 per cent per

[54] Caja de Credito Agrario, Industrial y Minero, *Informe de Gerencia*, December 31, 1966, p. 30.

[55] Field observations suggest that a return of 30 per cent on capital is not unusual.

[56] Caja de Credito Agrario, *op cit*., p. 41. The exchange rate is about 16 pesos to one US dollar.

annum.[57] Thus the Caja, instead of assisting the poor, deprives them of their savings and transfers its capital gratis to the more privileged members of the community.[58] Instead of encouraging householders to save and producers to economize on capital, the Caja encourages consumption and unemployment.

Table I:10

COLOMBIA: A SELECTION OF LOANS OF THE CAJA DE CREDITO AGRARIO

Object of Loan	Number of loans	Approximate average value of loan (pesos)
Crops grown by small farmers:		
Tobacco	6,443	2,250
Coffee	48,723	2,800
Manioc	14,471	2,280
Typical activities of large farmers:		
rubber	190	11,000
oil palms	239	7,900
cotton	2,507	19,600
beef cattle	104,251	6,300
pastures	12,750	5,300
irrigation equipment	829	18,000

Source: Caja de Credito Agrario, Industrial y Minero, *Informe de Gerencia,* December 31, 1966, Anexos.

The introduction of monopoly elements into the markets for land and capital, which we have described, and the consequent presence of monopsony power in the labour market implies that latifundistas and minifundistas will face an entirely different set of relative factor prices. This can be expressed in two ways, either as divergences of (actual or implicit) market prices from hypothetical competitive prices or as differences in factor price ratios.

Let 'i', 'r' and 'w' stand for the competitive price of capital (K), land (L) and labour (N), respectively; the subscripts 'l' and 'm'

[57] About the only way the poor can save without having their capital gradually consumed by inflation is by filling their teeth with gold.

[58] Industrialists can also obtain credit cheaply. 'Blue chip' manufacturing enterprises in Medellín pay about 12 per cent for long-term capital and 16 per cent for working capital.

stand for latifundia and minifundia. The market power of the former is such that

$$r_l < r < r_m;$$[59]
$$i_l < i < i_m; \text{ and}$$
$$w > w_l \geqslant w_m.$$[60]

Given these relative factor prices one would expect different factor proportions to prevail in large and small farms. Specifically, one would expect

$$\left(\frac{K}{N}\right)_l > \left(\frac{K}{N}\right)_m \text{ and}$$
$$\left(\frac{N}{L}\right)_l < \left(\frac{N}{L}\right)_m.$$[61]

In fact, this is what one finds. The large landowners tend to produce more capital intensive crops, e.g. irrigated cotton, and to use land rather wastefully, e.g. by grazing cattle on natural pastures. On the other hand, the minifundistas—facing extremely high prices of land and capital relative to the low implicit price of their own labour—tend to produce more labour intensive crops, e.g. foodgrains, and to have very high yields.[62] In other words, the operation of the price system influences both the composition of output and the techniques of production employed on the two types of farms: land being over-utilized on the small farms and consequently creating problems of erosion, while land is under-utilized on the large farms.

[59] In Colombia there is an additional factor that the land tax per ha. varies inversely with the size of the farm. (CIDA, *op. cit.*, p. 76.)

[60] Monopsony power reduces the wages of both the minifundista and the worker on a latifundium below what they otherwise would be, but w_l and w_m should be equal. In some cases, however, 'social laws', e.g. agricultural minimum wage legislation, if enforced, may increase w_l relative to w_m.

[61] The capital: land ratio is indeterminate.
That is, $\left(\frac{K}{L}\right)_l$ will exceed $\left(\frac{K}{L}\right)_m$ only if $(i/r)l < (i/r)m$.

It seems unlikely that this condition will hold in practice since capital markets seem to be less imperfect than the market for land. Thus it is likely that $\left(\frac{K}{L}\right)_l < \left(\frac{K}{L}\right)_m$.
This hypothesis appears to be consistent with the limited data available on Colombia, which show not only that the minifundia have a higher density of tree crops and livestock per ha. than the latifundia; they also have a higher density of machinery per ha. (CIDA, *op. cit.*, p. 167)

[62] In Colombia yields per ha. on the minifundia are fourteen times those of the large latifundia. (CIDA, *op. cit.*, p. 171.)

4 POLICIES AND A PREDICTION

We have seen that the latifundia system is both economically inefficient and socially unjust. The under-privileged groups in Spanish America appear to have little political power,[63] are pushed to the margin of the economy as consumers and are strongly—and unfavourably for them—integrated into the economy as sellers of commodities and their labour. This situation will persist as long as political power, land and capital are concentrated in the hands of a few families. If one is determined to increase equity and raise efficiency, groups with common interests will have to be organized and there must be established a counter-force to the long-established vested interests of society.

In some cases relatively 'small' changes could have a substantial impact upon the bargaining power and living standards of the rural poor. In Ecuador, for example, the rural population in the Sierra is 1·5 million. The social and economic position of these people would greatly improve if the government would take two steps:

(i) declare all roads, whether on private haciendas or not, open to free public access, and
(ii) nationalize the *páramos*, mountainous regions at the 3,000–4,000 meter level, and allow persons who own less than 15 hectares to gather wood and use the pasture freely.[64]

As the *páramos* are uncultivated wastelands, neither of these measures would directly reduce the *hacendado's* control over productive resources. Both of them, however, would reduce his monopoly of the land and his consequent control over the labour market. At present, the *huasipunguero* earns roughly $150 per year; 32 per cent of this comes from his agricultural activities, 52 per cent from livestock, and only 16 per cent as wages from working on the hacienda —although he spends at least 40 per cent of his time there.[65]

Thus the advantages of the proposed minor changes are that they would

[63] UN, ECLA, *El Desarrollo social de America Latina en la postguerra*, p. 35, states, '. . . *las areas rurales de America Latina . . . permanecieron relativament segregadas del poder central*', and '*existen . . . indicios de que las masas rurales no llegaron a constituir, en general, un sector electorál apreciable*'.

[64] Depending upon the definition applied, anywhere from 500,000 to 1·2 million ha. (20–40 per cent of all the land in the Sierra) would be subject to nationalization.

[65] R. Baraona, *op. cit.*, p. 320.

(i) almost completely liberate the *yanapero* from the necessity of providing labour services gratis;
(ii) reduce the *huasipunguero's* dependence on the hacienda's grazing lands and thereby liberate the major portion of his income from hacienda control, and
(iii) increase the bargaining power of the workers in the labour market and hence increase their wages.

Even such simple changes, however, are likely to be strongly resisted by the landowning class, precisely because it would undermine the props upon which their wealth and political power are built. Thus, on the one hand, these measures would be firmly opposed by the dominant political forces; on the other hand, they would not represent a complete solution to the problem of development and social transformation in Spanish America. As partial measures would be largely unsuccessful in achieving the larger objectives and would be bitterly resented anyway, there is little to lose in pressing for more radical and satisfactory solutions.

The essential ingredient in any programme of development and social transformation is a fundamental change in agrarian institutions. The expropriation and redistribution of land and the organization of a mixed system of producer cooperatives, family holdings[66] and state farms should be the basis for all subsequent action in this field.

Tenure reforms, however, are only the first step in the process of social transformation. It is important, therefore, that this step be accomplished swiftly and cheaply. This cannot be done if compensation of the landlords is based upon the commercial value of the expropriated land. Land values in Spanish America are much higher than one would guess knowing only the productivity of the soil. This is because the commercial value includes

(i) the value of land as an item of social prestige, and
(ii) the value of land as a hedge against inflation. The inflation itself, however, is caused in part by the failure of agriculture

[66] Many economists from the United States are unduly biased in favour of the family farm system. This certainly is not a general solution to the social problem in Spanish America and under some circumstances small, individual holdings may ultimately be prejudicial to the interests of the intended beneficiaries. R. E. Christ, *op. cit.*, p. 140, points out that '. . . at present, whenever the land of a community is divided up among the Indians and given to them in fee simple, it usually comes to be owned within a short time by the local *hacendado*, or landowner, who, with an eye to business, is likely to be the first to favour the breaking up of community holdings and the giving of "full civil rights" to the Indians'.

to expand production and the unwillingness of the upper classes to pay their taxes.[67]

(iii) Finally, market prices reflect the capitalized value of the right to exploit labour. One does not buy just agricultural land; one buys an entire enterprise, including its monopoly power.

Compensation should be based upon the value of the land, exclusive of these three factors, as operated under current techniques of production. Only on this basis, for example, would it be reasonable and just to require the new owners of any family farms that might be created to pay the full purchase price of the land acquired under the reform.

Tenure reforms will have to be supplemented by the creation of agricultural labour unions, the provision of a minimum agricultural wage, and strict legal enforcement. Marketing, storage, and transport facilities will also have to be provided. Perhaps the best way to do so would be through a system of national marketing co-operatives which compete with the existing intermediaries and which guarantee minimum prices. Similar institutions must be responsible for providing credit and other important inputs, e.g. fertilizers. Rural education will have to be expanded and adapted to the new conditions. Finally, once a land reform has ensured that those who work receive the fruits of their efforts, the unemployed should be mobilized in a programme of rural public works.[68] Such a programme of rural public works would be the quickest way to provide full employment, increase investment and transform the countryside—and thereby guarantee that an eventual increase in consumption would be possible.

One would hope that the above policies would be obvious and command general acceptance, yet even the rather progressive Chilean government had an entirely different conception of *promoción popular*.[69] In a major speech shortly after his election President

[67] Conditions in Peru—and much of Spanish America—are summarized succinctly by two articles which appeared on the front page of *La Prensa* (Lima) on February 18, 1965. One article reports an invasion of agricultural land in the District of Paca while the other reports that '*Solo 25 mil Personas Pagan sus Impuestos*.'

[68] See T. Balogh, 'Agriculture and Economic Development', *Oxford Economic Papers*, February 1961.

[69] Many of the above recommendations *have* been endorsed by Latin American intellectuals. See, for example, J. Chonchol, '*El Desarrollo de America Latina y la Reforma Agraria*', *Curso de Capacitacion de Profesionales en Reforms Agraria*, Vol. I. Mr Chonchol was one of President Frei's major advisers on agrarian reform. At the beginning of the new administration it was hoped that 100,000 families would benefit from the land reform by 1970. This target has now been reduced to 40,000 and it will be surprising if in fact many more than 20,000 families benefit. Mr Chonchol resigned his post in 1969 in protest.

Frei presented his plan for *promoción popular* and announced that a new ministry would be created to execute it.[70] The plan was chiefly concerned with increasing welfare measures and consumption of the low income urban classes *'que forman un cinturón de miserias en torno a nuestras grandes ciudades.'* Specifically, the government hoped to do such things as pave 29,000 meters of sidewalks and streets; provide water, garbage disposal and telephone facilities to thousands of urban dwellers; construct 50 social centres and numerous parks and athletic fields, etc.

All of these measures, of course, are defensible; they cater to pressing needs and will help alleviate the misery of an important segment of the population. Still, one must question whether expenditure in this direction represents a good allocation of resources. All of these projects are essentially welfare measures; they increase the consumption of the urban masses, but they do not generate much additional employment, nor do they represent investment which will permanently increase the productive capacity of the economy.

The Peruvian scheme of *Cooperacion Popular* has a slightly different orientation. The urban masses are not completely neglected but the thrust of the programme is towards giving '*prioridad dentro del as acciones para el desarrollo a las de caracter infraestructural.* . . .'[71] Everything from airports to churches and irrigation canals have been financed under the scheme. The emphasis, however, has been on constructing provincial roads and schools. These are measures which increase labour mobility and skills, but they do not directly increase productivity.

The alternative to increasing urban consumption or investment in infra-structure is to increase directly productive rural investment. It has been conclusively demonstrated in several countries that properly organized labour-intensive rural investments (a) are an excellent way to mobilize the masses for development, (b) are inexpensive, (c) can have a very short gestation period, and (d) provide large returns on capital expenditure.[72] The author has personal experience in Algeria

[70] The text of President Frei's speech can be found in *El Mercurio* (Santiago, Chile), December 11, 1964.

[71] Comision Ejecutiva Interministerial de Cooperacion Popular, *op. cit.*, p. 57; emphasis in the original.

[72] See, for example, R. Gilbert, 'The Works Programme in East Pakistan', *International Labour Review*, March 1964; K. B. Griffin, 'Algerian Agriculture in Transition', *Bulletin* of the Oxford University Institute of Economics and Statistics, November 1965; E. Costa, 'Manpower Mobilization for Economic Development in Tunisia', *International Labour Review*, January 1966.

with projects of the type recommended. In the Tizi-Ouzou region, for example, working in poor and eroded soils, the benefit-cost ratio on investment in the first year was a minimum of 1·51. Spanish America can neglect this experience only at her peril.

A Short-run Prediction

Yet she probably will neglect it. Historically, the Spanish American middle classes have tended to imitate their social and economic superiors and to attempt only to force the latter to allow them to participate in the privileges of power.[73] This coalition then has been able to fragment the lower classes by granting partial concessions which are not distributed evenly. This formula, particularly as applied in Chile, has been successful in avoiding a social revolution and in maintaining essentially untouched the privileges of the wealthy. Thus the middle-class reforms in voting rights, labour legislation, social security and education were incomplete and fragmentary; they usually were confined only to certain sectors of the urban population and seldom were effectively enforced.[74] Furthermore, in so far as there was a change in the distribution of income, it was the urban middle class and unionized labourers who gained—not at the expense of the rich—but at the expense of the rural masses, *Lumpenproletariat* and unorganized workers.[75] The increased welfare services were financed not by higher or more efficient direct taxation of the upper income groups, but on the contrary, through a combination of inflation and additional indirect taxation.[76]

The *campesino* has been systematically ignored by almost all the progressive parties in Spanish America. The latter have chosen to direct themselves to the organized and privileged *élites*, e.g. the unionized industrial workers, students, intellectuals, and bureau-

[73] The way by which the middle class was able to force concessions from the aristocracy is explained by ECLA, *El desarrollo social de America Latina en la postguerra*, p. 102, as follows: '*En la fase ascendente del proceso político los sectores medios iniciaron su acceso al poder apoyándose por lo gėneral en las masas obreras, y creando, en consecuencia, diversas instituciones cuyo propósito querido o "manifiesto" fue la mejora del* status *social y economico de empleados y obreros. Pero el efecto no declarado o "latente" de esas instituciones parecería haber consistido mas bien en la expansion y mejoramiento de las capas medias mismas.*'

[74] *Ibid.*, pp. 113–14.

[75] In Chile, for instance, the workers' share of the national income declined from 33·7 per cent in 1940 to 26·5 per cent in 1960.

[76] See O. Sunkel, 'Change and Frustration in Chile', in C. Veliz, ed., *Obstacles to Change in Latin America*, pp. 132–3. Indirect taxes in Chile rose from 54 per cent of the total in 1940 to 64 per cent in 1960.

crats.[77] These groups, first, represent a small proportion of the population in most countries, and second, rarely are interested in a profound transformation of society; but instead are concerned primarily with widening their own opportunities for advancement. The programmes recommended in this chapter are directed to the benefit of the rural masses and are much more radical than the essentially 'welfare-state' measures usually sponsored by the urban middle class. Thus experience and logic would lead one to be a little sceptical whether they will be introduced by Spanish America's urban-based, middle-class parties.

[77] Frantz Fanon believes that this is true in the majority of undeveloped regions. See *Les damnés de la terre*.

CHAPTER II

Resource Transformation and Foreign Trade

In the first chapter we discussed the social and economic structure of the Spanish American countries. In that discussion we were primarily concerned with the inter-relationships between regions and groups in a single economy; external or foreign relationships were largely ignored. The remaining chapters of this study will focus upon various aspects of the latter problem. We shall begin in this chapter by considering, first, the composition of visible trade; secondly, the transformation problem and static allocation analysis; and finally, fluctuations and trends of export prices and movements of the terms of trade.

1 THE STRUCTURE OF TRADE

Like most underdeveloped countries in all regions of the world, the foreign trade sector is extremely important to the Spanish American economies. The role and character of this sector can quickly be described: (i) exports plus imports represent a high proportion of national income; (ii) exports usually are highly specialized in only one or two products and several countries are virtually mono-exporters; (iii) major exports are invariably either primary agricultural commodities or minerals. As a result the economies of Spanish America are heavily dependent upon a few major primary products and they are highly vulnerable to changes in international demand and prices. The extent of dependence on foreign trade is illustrated in the table below.

As can be seen, the foreign trade sector represents between 25 and 60 per cent of the national income of all the Spanish American countries except Argentina. Only Peru—which exports several minerals, cotton, sugar and fish meal—has a truly diversified export sector. Argentina, which might appear from the table to be diversified, is in fact highly specialized in the export of meat, cereals and wool.

Table II:1

TRADE DEPENDENCE

	Exports plus Imports as per cent of national income	Major Export	Major Export as per cent of all exports
Argentina	16	meat	26
Bolivia	53	tin	85
Chile	26	copper	73
Colombia	34	coffee	64
Ecuador	39	bananas	61
Paraguay	35	meat	36
Peru	51	copper	27
Uruguay	27	wool	45
Venezuela	62	petroleum	92

Source: IMF, *International Financial Statistics.* See also OAS, *Latin America: Problems and Perspective of Economic Development, 1963–64,* 1966, Table 1–5 and, for the early post-war period, D. H. Pollock, *International Economic Instability.*

Table II:1, which is dramatic enough, tends to understate the dependence of the nine economies upon foreign trade. This is because it presents only a static picture of the relative importance of a few export products. The most important distinction between a Spanish American economy and a Western industrialized nation, however, is not in the size of the foreign trade sector but in the behaviour of domestic investment. In contrast to Western European, British and North American experience, investment in Spanish America is an endogenous variable, i.e. the rate and level of investment is determined largely by production and marketing conditions for exports and the consequent capacity to import. This is particularly true in countries, first, which have no domestic capital goods sector and hence must obtain their machines from abroad and, second, in countries which have import-substituted a large number of consumer goods industries and hence any decline in foreign exchange receipts forces the country to reduce its imports of capital goods.

Several Spanish American countries already have replaced a large number of imported consumer goods with domestic substitutes. In these countries the great majority of imports consist of raw materials and investment goods. In Argentina, for instance, about 92 per cent of all imports are raw materials, fuels and capital goods. The proportion is almost as high in Chile (84 per cent) and Uruguay (83 per cent). In Peru, consumer goods account for 24 per cent of all imports:

10 per cent are durable consumer goods (automobiles, television sets, etc.) and 14 per cent are non-durables (e.g. food). In countries such as these it is rather difficult to reduce imports of consumer goods much more, and consequently any reduction in the availability of foreign exchange tends to lead to a reduction in imported capital goods and the level of investment.

The relationship between exports and investment is unusually clear in Venezuela. During 1950–58 exports plus tourism increased at an annual rate of 7·8 per cent and investment grew at 7·9 per cent. During 1958–64 the rate of growth of exports plus tourism declined to 4·6 per cent and the rate of investment consequently plunged to −2·3 per cent per annum. A similar relationship—this time between imports and investment—has been observed in Colombia: a one per cent change in imports is associated with a 1·3 per cent change in investment.[1] Thus, not only does the behaviour of the foreign trade sector determine the level of national income, it also—through its link with domestic investment decisions—influences the general rhythm of economic expansion and stagnation.

Not only are exports usually highly concentrated on one or two products, they also are dependent upon a relatively few foreign buyers. Spanish America has failed not only to diversify her production, she has also failed to diversify her markets. What is true of our nine countries is true too of Latin America as a whole. Roughly 35 per cent of Latin America's exports are sold in North America. About a third are sold in Western Europe: 21 per cent in the European Economic Community and about 11 per cent in the European Free Trade Area. Between three-fourths and four-fifths of the region's exports are sold to the industrial nations. Less than 10 per cent are sold in other Latin American countries, and trade with Asia, Africa and the Communist countries is negligible.[2]

This pattern is repeated in the nine Spanish American countries, as can be seen in Table II:2. Two-thirds of Spanish America's exports are absorbed by the Western industrial nations. Sales to the Communist countries, the rest of Latin America and the other underdeveloped countries are, in most instances, rather slight.

This dependence on only a few markets means that some of the Spanish American economies are sensitive to changes in the economic

[1] K. G. Griffin, 'Coffee and the Economic Development of Colombia', *Bulletin of the Oxford University Institute of Economics and Statistics*, May 1968.

[2] For detailed data see ECLA, *Economic Bulletin for Latin America*, Vol. VII, No. 2, October 1962, p. 157 and *Economic Survey of Latin America 1964*, 1966, p. 196.

performance of their major customers. The behaviour of the US economy in particular (which absorbs about 30 per cent of Spanish America's exports) is of vital importance. The cyclical instability and slow growth which until recently has characterized that economy has constituted a disturbance to Spanish America. In the US recession of 1948–49 GNP declined only 0·1 per cent, but US imports from Latin America declined 2·2 per cent. Similarly, in the recessions of 1953–54 and 1957–58 GNP declined by only 1·6 and 1·7 respectively, whereas imports from Latin America declined by 4·4 and 4·7 per cent respectively. Those countries which export minerals can be particularly hard hit. For example, Bolivia's exports declined by 20·2 per cent in 1953–54, and by 33·6 per cent in 1957–58.

Table II:2

DESTINATION OF SPANISH AMERICAN EXPORTS, 1965
(per cent)

	U.S.A.	Latin America	Western Europe	Other
Argentina	6	17	59	18
Bolivia	42	2	53	3
Chile	31	8	48	13
Colombia	47	6	37	10
Ecuador	59	8	30	3
Paraguay	25	30	30	15
Peru	34	9	43	14
Uruguay	17	8	64	11
Venezuela	35	8	19	38
Spanish America	29	10	37	24

Source: US Department of Commerce.

Note: Venezuela's exports to 'Other' nations are principally exports of crude petroleum to the Netherlands Antilles for refining.

Events in Europe can also be very disturbing. The repeated balance of payments problems in the UK have tended to slow down the growth rate of this important market. The creation of preferences by the EEC in favour of imports of tropical products from the ex-French Empire, and particularly the new African countries, has worked to the disadvantage of Spanish America. Italy's agreement with Somalia to import bananas may have had a similar effect.

The problem of preferences and discrimination is more general and more serious than the above comments would seem to imply.

Not only do some countries—particularly the EEC—discriminate in favour of some tropical producers and against others, but *all* the countries of the industrialized West discriminate in favour of imports of primary products and against imports of manufactured goods. This discrimination is achieved through the tariff structure. For instance, the tariff on raw cotton in the US, UK, and EEC is zero; the tariff on yarn is 14, 8, and 8 per cent respectively; while the duty on woven fabrics is 20, 28 and 17 per cent. This structure of tariffs in the major importing countries tends to force Peru, for example, to remain an exporter of raw cotton and tends to inhibit the development of textile manufactures.

These duties are cumulative, since they are levied on the total value of an imported product and not just on the value added. Hence the degree of protection afforded to the manufacturer in the industralized nation is considerably greater than the apparent tariff rate. For example, assume a manufactured product is imported at an official tariff rate of 15 per cent. If we further assume that this product is made up half of a raw material which could have entered free of tariff and half value added, then the effective degree of protection is not 15 per cent but rather 30 per cent! The tariff applies not only to the value added in manufacturing but also to the raw materials embodied. (We shall return to this topic in Chapter VI.)

Imports of the Spanish American economies tend to be the mirror image of their exports. Most of the countries have managed to

Table II:3

SOURCES OF SPANISH AMERICAN IMPORTS, 1965
(per cent)

	U.S.A.	Latin America	Western Europe	Other
Argentina	23	26	39	12
Bolivia	44	12	28	16
Chile	39	23	31	7
Colombia	48	11	32	9
Ecuador	41	12	35	12
Paraguay	23	20	36	21
Peru	40	12	32	16
Uruguay	13	30	40	17
Venezuela	51	3	30	16
Spanish America	39	15	33	13

Source: US Department of Commerce.

establish manufacturing enterprises which produce light consumer goods for domestic consumption. Relatively few have been able to begin producing heavy industrial goods. Most, however, are self-sufficient (or nearly so) in basic foodstuffs such as cereals and potatoes. Almost everything else must be imported. Even pasteurized milk was at one time regularly imported by Venezuela from the United States.

The major imports, as one might expect, consist of the more sophisticated products, e.g. consumer durables and industrial machinery, and industrial raw materials. These tend to come from the industrialized nations of the West, and particularly from the United States. Similarly, services such as shipping usually are provided by the advanced economies. Indeed, Argentina, Ecuador and Venezuela tend to have a surplus on merchandise account and the other countries usually have only a small deficit; it is the large imbalance on the services account which is directly responsible for most of the balance of payments problems in the region. In general, almost 40 per cent of Spanish America's imports originate in the United States, although some countries, e.g. Venezuela and Colombia, are much more dependent on US supplies. Imports from Western Europe constitute over 30 per cent of the total and Argentina and Uruguay in particular are dependent on this source.

Table II:4

FOREIGN TRADE BALANCE, 1965
(millions of US dollars)

	Exports	Imports	Balance
Argentina	1,493	1,199	294
Bolivia	110	126	−16
Chile	685	604	81
Colombia	537	454	83
Ecuador	161	171	−10
Paraguay	57	52	5
Peru	666	719	−53
Uruguay	191	150	41
Venezuela	2,744	1,453	1,291
Spanish America	6,644	4,928	1,716

Source: International Monetary Fund.

Since the end of World War II there has been a tendency in the more industrialized Spanish American countries, i.e. in all except Bolivia and Paraguay, for imports of raw materials and semi-manufactures to grow considerably faster than imports of finished goods. There is a parallel tendency for imports of consumer goods to increase more rapidly than imports of capital goods.

2 THE TRANSFORMATION PROBLEM

Static Allocation Analysis

Standard international trade theory usually illustrates the production possibilities of a country by referring to a transformation curve or production frontier. This curve indicates all the possible combinations of any two goods, say, tin (T) and potatoes (P), that could be produced with existing resources and technology. The transformation curve, in turn, is derived from a contract curve (i.e. the locus of points of tangency of iso-product curves) inscribed in a box diagram whose axes represent the fixed quantities of the (assumed) two factors of production. The two factors of production, furthermore, are assumed to be freely substitutable for each other (so that the isoquants will be smooth and continuous) and not specific to any particular industry (so that factors can be freely transferred from one sector to another).

The transformation curve so derived is drawn concave to the origin. This is combined with a hypothetical construction called a community indifference curve—which reflects the pattern of demand as determined by consumers' tastes and the distribution of income. Under conditions of autarky resources are optimally allocated and welfare maximized where the community indifference curve is tangent to the transformation curve.[3] This is point Q on the diagram below. The relative prices of P and T are represented by the slope of the tangent at Q. At this point T_0 tin and P_0 potatoes are produced and consumed.

It can easily be shown that trade will take place between two countries whenever the cost (or price) ratios differ. These ratios may differ because natural resources vary, because the population is unequally distributed, because of differences in human capacities and abilities, because historical accidents have influenced the volume and composition of the existing stock of capital, because specific factors are unequally spread throughout the world, or because tastes differ. In other words, trade will occur if the transformation curves are the same but tastes differ, or if tastes are similar but the production pos-

[3] For the derivation of the transformation curve see T. Scitovsky, *Welfare and Competition*, Ch. VIII; for its application to international trade see C. P. Kindleberger, *International Economics*, Chapters 5 and 6.

sibility curves are different, or if both supply and demand conditions are dissimilar.

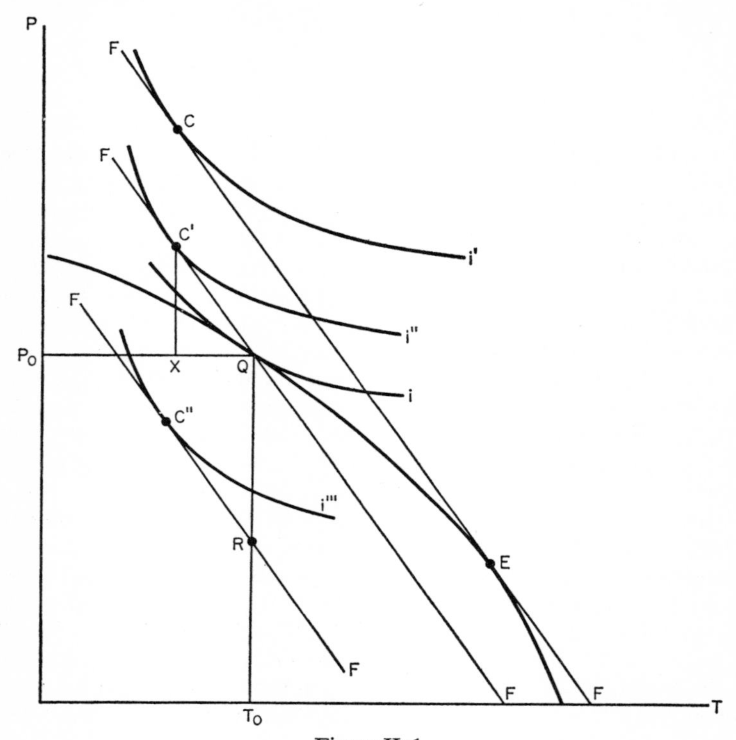

Figure II:1

The important normative implication of the above analysis is that free trade will make a country 'better off', in the sense that more goods and services will be available under free trade than under autarky. Thus, in figure II:1, if a country faces a free trade price ratio (FF) different from its autarkic price ratio, it can profitably specialize more in the production of tin and exchange this for potatoes. In doing so it would move to a higher indifference curve i′. More tin would be produced and more potatoes consumed: production would take place at E and consumption would take place at C.

These results would have to be modified, however, if the assumptions upon which they are based were changed.

In many of the Spanish American countries large numbers of

resources are specific to a particular industry. Venezuela cannot convert her petroleum deposits into dairy farms; Chile cannot convert copper ore into furniture; Bolivia cannot transform tin mines into potatoes. Thus the transformation curve becomes a rigid production point. Moreover, Ecuador and Colombia cannot easily and quickly expand their banana and coffee plantation sectors or convert them to other crops without undertaking large investments which, in some cases, have long gestation periods. This immobility of factors of production due to specific natural resource endowments is reinforced by the low level of investment and expenditure on education which characterize the region. Once a particular structure of production has been selected, for whatever reason, labour and capital tend to become immobile because they become incorporated into the productive process in the form of durable and specific skills and equipment. This stock of labour skills and capital equipment cannot readily be transferred from one industry to another. In the absence of adequate facilities for training labour and a high level of investment a country has little flexibility or choice in determining the composition of output. Both the shape and position of the transformation curve depend upon the volume of resources devoted to growth. Hence poor and stagnant economies invariably are subject to inertia and have rigid production possibility schedules.[4]

The implications of this rigidity are illustrated in our figure.[5]

It is commonly supposed that the gains from international trade arise as a result of opportunities for specialization. Actually, some of the gains are due to opportunities for *exchange* of goods at price ratios different from those prevailing under autarky; specialization is not essential. This can be seen in Figure II:1. If we assume that factors of production are immobile we obtain a single point transformation curve, Q, at which production and consumption initially take place. Under the free trade price ratio of FF production remains at Q—there is no specialization—but consumption moves to C′ on a higher

[4] In other words there may be both static increasing returns due to economies of scale and dynamic increasing returns associated with the rate of capital accumulation. Neither of these will be exploited, however, if the level and composition of development effort are inadequate. A single-point transformation curve is largely an historical phenomenon; it tells us nothing about the direction in which development should proceed in the future nor about the returns one could expect from additional effort

[5] Also see G. Haberler, 'Some Problems in the Pure Theory of International Trade', *Economic Journal*, June 1950; S. B. Linder, *An Essay on Trade and Transformation*, pp. 24–8; J. Bhagwati, 'The Theory of Comparative Advantage in the Context of Underdevelopment and Growth', *Pakistan Development Review*, Autumn 1962.

indifference curve. The country exports QX of T and imports XC′ of P. Note, however, that the inability to transform resources reduces (although it does not eliminate) the gains from trade. Had the country been able to specialize, for example, at E, the community could have reached an even higher indifference curve i′.

The assumption in the above paragraph that factor prices are flexible is very important. If, in addition to factor immobility, real factor prices are also inflexible, there may be a loss from international trade due to a decline in the output of one of the commodities. If the price of one good falls, the prices of factors used in the production of that good must fall. If they do not, the output of the commodity must fall. Just how far production declines depends upon cost conditions in the industry. If, for example, production occurs under conditions of rising marginal cost, then as output contracts, costs will fall, and a point may be reached where the industry is once again competitive.

This too is illustrated in the diagram. The relative price of P has fallen after the introduction of free trade. Factor prices were unable to adjust so that output of P fell from Q to R. Consumption now takes place at point C″ on a lower indifference curve, i‴. There has been a loss from trade.

Unfortunately, the Spanish American economies exhibit some of the features of the above models. We have already argued that specific natural resource endowments, inadequate educational and training facilities, and low domestic savings ratios have created transformation problems for these nations. It is also true that real factor prices, notably wages and salaries, tend to be inflexible downwards. Part of the explanation for this is that large numbers of people are living at 'subsistence' levels of income. This is particularly true of Indians, small peasants, and rural workers. On the other hand the subsistence income of other groups, e.g. miners and industrial workers, does not correspond to a physiological minimum, but rather to a politically determined minimum level of consumption. The common use of the expression '*sueldo vital*' attests to the importance of this phenomenon. A third reason why wages are sticky (even if not rigid) downwards is that small but important groups such as professionals and government employees are well organized and are able to exert considerable bargaining strength to prevent a fall in their real incomes.

This rigidity of real, and not just money, incomes has important implications not only for an analysis of the gains from foreign trade, but in other areas as well. For example, devaluation almost inevitably leads to pronounced inflation and hence tends to be self-defeating; price increases induce compensating wage increases and hence inflation tends to be cumulative. These matters are discussed in some detail in Chapter V.

The Beneficiaries of Trade

So far we have shown that in economies characterized by transformation problems and institutional rigidities the gains from trade, though perhaps not negative, are considerably smaller than one might normally presume. Even if one accepts, however, that international trade is likely to lead to a positive but static, once-for-all increase in national income, this may not be entirely relevant in fragmented economies of the type we are considering. The major preoccupation of the Spanish American countries is not with the level of their national income, but with its rate of growth and equitable distribution. From this point of view the way in which the gains from trade are utilized may be much more important than the static magnitude of the gains. Hence it is essential that the groups in society which reap the major benefits from trade be analysed.

The major direct beneficiaries can be catalogued roughly into two groups. The first group consists of foreign owned enterprises in the export sector. They are prominently located in the petroleum and mineral industries in Chile, Peru, Venezuela and elsewhere. The immediate advantages of additional foreign trade in these activities accrue to the foreign enterprises in the form of increased profits, which frequently are repatriated. The impact of foreign capital and foreign capitalists on Spanish America, however, is a complex topic and it is discussed in considerable detail in the next two chapters. All that should be noted here is that there are two secondary beneficiaries of foreign economic activity, *viz.* the government, which can impose high taxes, and the workers of the enterprise, which usually receive wages markedly higher than the average prevailing in locally owned firms. As a result, both the government and a small minority of privileged workers acquire a strong vested interest in perpetuating the existing pattern of trade. The benefits to the rest of the economy may be very slight, especially if the government uses its revenues to expand the bureacracy rather than to increase investment.

Owners of large tracts of land producing export crops constitute the second important group which currently benefits directly from the existing structure of foreign trade. The banana growers of Ecuador and the *estancieros* of Argentina are typical examples. Only a small proportion of the profits received by the landowning classes are normally invested in the Spanish American economies. Tax rates in the agricultural sector are low and tax evasion is widespread.[6] The marginal propensity to consume of these groups is high and the import component of consumption—in the form of consumer durables and foreign travel—is substantial.[7] The remaining part of the surplus frequently flees abroad and is deposited in foreign banks.[8] Thus high profit rates from exports of agricultural commodities may not always be translated into high and sustained rates of domestic investment.

We must recognize that in economies in which markets are fragmented and political authority is weak specialization in the production of those commodities in which the static gains are greatest may be less desirable than specialization in other commodities. This is because it may be easier to convert the surplus into investment in the latter and, hence, accelerate the rate of growth. Thus the question as to who receives the gains from trade—whether landowners, foreigners, industrialists, workers, or the government—is important for dynamic analysis. This importance is increased (i) the more imperfect is the capital market, (ii) the greater are the variations in the propensities to consume between socio-economic groups, and (iii) the easier it is to evade taxes in one sector compared to another.

The argument can be expressed in algebraic terms. Suppose a country has a choice of investing a given amount either in cotton which yields a profit (exclusive of monopoly elements) of Pc, or in a fruit juice factory with a profit of Pf. Assume $Pc > Pf$. Assume also, however, that the savings habits of industrialists and agricultural producers are such that, in combination with the sectoral tax rates and possibilities for evasion, the proportion of Pf available for reinvestment (α) is greater than the proportion of public and private savings (β) which can be extracted from Pc. In these circumstances

[6] See J. Strasma, '*Técnicas de evaluación y planificacion fiscal*', in K. B. Griffin and E. Garcia, eds, *Ensayos Sobre Planificación*, Santiago, 1967.

[7] See M. J. Sternberg, *Chilean Land Tenure and Land Reform*, Ph.D. thesis, University of California, 1962.

[8] It is, of course, impossible to obtain precise estimates of the outflow of private capital from Latin America. Many economists believe there is about $10 thousand million of Latin American capital abroad. Data collected by the IMF suggest that the minimum estimate of the outflow during the period 1952–63 was $4·1 thousand million (see P. Host-Madsen, 'How Much Capital Flight from Developing Countries', *Finance and Development*, March 1965). Robert Triffin, using ECLA data, suggests that the total net outflow of Latin American capital during the eight years 1956–63 was between $5 and $3 thousand million. (See R. Triffin, 'International Monetary Arrangements, Capital Markets and Economic Integration in Latin America', *Journal of Common Market Studies*, October 1965, pp. 74–6.)

the rate of growth would accelerate if the country specialized in producing fruit juice—even though cotton is more profitable—provided the difference in the savings ratios is greater than the difference in profitability, that is, provided $\alpha/\beta > Pc/Pf$ or $\alpha Pf > \beta Pc$.

Thus, in general, investment and growth will increase if exports expand, but the amount of the increase will depend upon the characteristics of the sector in which the initial expansion occurs.

Trade and the Transfer of Knowledge

It frequently is stressed that international trade is an important vehicle for transferring knowledge. Those who make such assertions usually imply that such a transfer is beneficial to the receiving countries. In general there is no doubt that this is true. But it is also true that not all transfers of knowledge are useful. Just as international commerce served as a vehicle for introducing European diseases to societies which were unable to resist them—and hence led to the decimation of large populations, so too can international trade serve as a vehicle for introducing Western ideas which are ill adapted to or inappropriate for the prevailing conditions in the host countries. Although on balance the exchange of knowledge may be beneficial to the world community, one cannot presume that there are no negative effects of such an exchange or that the negative effects are evenly distributed. In fact, there are reasons to believe that trade in ideas may be systematically biased against the underdeveloped countries, in the sense that the importation of foreign ideas may induce a reaction in the host country which is inimical to economic growth. In any particular case the negative bias may or may not be over-compensated by the more diffused positive effects.

More specifically, ideas transmitted by foreign trade help to determine (a) what is consumed, (b) what non-traded goods are produced, and (c) the techniques by which goods are produced. The term 'international demonstration effect' has been used to describe the first of these influences, but since the problem is perfectly general, we will use the same term to describe the other two cases as well.

We have shown above that international trade, by making available possibilities for exchange, *ceteris paribus*, enables a consumer to move to a preferred position, i.e. on to a higher indifference curve, and thereby satisfy his given wants more perfectly. In addition, the process of trade itself assists the transfer of information and ideas and, consequently, changes the environment in which the given wants were formulated. The net result of interchange, therefore, is to alter the taste patterns of the community. New desires are created;

new ambitions for consumption are encouraged. These new ambitions can only be satisfied by additional imports and higher incomes. Thus the effect of trade is to widen the 'aspiration-income gap'[9] (and to invalidate the static, *ceteris paribus* analysis). This widening of the gap is reflected not only in an increased demand for imported consumer goods but also in demands for a rapidly rising level of consumption in general. Consumers in the less developed economies try to imitate the standard of living of the wealthy nations.

This phenomenon is particularly acute in Spanish America because of the proximity to and close contact with the United States. High aspirations are generated and sustained by constant exposure through mass communications media to the extremely high standard of living of the Northern Neighbour. This constant pressure to increase consumption falls in part on the balance of payments and in part on the savings effort of the community. The demonstration effect leads to a non-reversible shift of tastes and a preference for present (as opposed to future) consumption. It is reflected in a balance of payments deficit and a low savings rate. Thus precisely during the period when growth becomes more urgent the means of achieving growth—savings and foreign exchange—become more scarce. To the extent (a) that it is the high income groups that import most of the (luxury) consumer goods and (b) that scarcity of foreign exchange is the principal bottle-neck restraining growth, a more equal distribution of income—by switching demand to home-produced consumer goods and releasing foreign exchange—would enable output to increase at a faster rate.

The second aspect we must consider is the influence of international trade upon institutions. In recent years it has become clear that not only is the demonstration effect working to raise consumption aspiration, it is also fostering institutions in Spanish America which are more appropriate to the welfare state. These measures are expensive and consumption-inducing; they use funds which could be used to expand productive equipment; they tend to lower savings and retard the rate of growth. Yet virtually every Spanish American country, no matter how poor, has begun to proliferate social security schemes, health services, pension plans, and programmes for subsidized consumption. Uruguay is furthest advanced in this direction, but Chile and Argentina are close behind, and the rest are gaining rapidly.

Welfare measures have nearly become a political status symbol

[9] See R. Weckstein, 'Welfare Criteria and Changing Tastes', *American Economic Review*, March 1962.

and it is considered reactionary to oppose them. Yet while public consumption expenditure has accelerated, expenditure on capital formation and vocational training has remained small. The consequences for growth of this distortion in the composition of government expenditure should not be underestimated. Government revenues are typically inelastic; population growth is very rapid. As a result an ever larger proportion of the government budget will have to be devoted to welfare services, and a smaller and smaller portion will be available for expenditure on education[10] and state investment in infrastructure and directly productive activities. The imitation of European institutions has led to a distortion of the composition of output in favour of public consumption and welfare services and against public investment, broadly defined. In the short-run, as discussed in the previous chapter, there may be political advantages in encouraging this trend, but in the long-run such policies will lead to a slower rate of growth—unless government revenues grow substantially. The Argentine case is an extreme example of what we have in mind. Between 1950 and 1963 inflation was rampant and the government experienced chronic and growing deficits. Meanwhile total government expenditure as a proportion of GNP fell from 15 to 13 per cent—and the fall in state investment was proportionately greater than the fall in current expenditure. Thus in Argentina not only did the composition of government expenditure become less desirable but the level of expenditure relative to GNP declined as well.

The final aspect of the demonstration effect that we will consider is its influence on the techniques of production. In Chapter V we argue that there are many good reasons why techniques of production in Spanish America should be relatively capital intensive, but the degree of capital intensity is often carried to an extreme. This is due, in our opinion, to a particular feature of the demonstration effect, *viz.* the importation of techniques of production that are highly capital intensive and labour saving by countries that are characterized by capital scarcity and unemployed labour. Imitation of the techniques used in industrialized economies may have two general effects upon underdeveloped countries such as those in Spanish America. First, it may accentuate an already severe unemployment

[10] There is a further problem that money is being spent on the wrong kind of education. Spanish America has adopted European attitutes toward education and as a result excessive emphasis is placed on literary, legal and humanistic training—at the expense of instruction in science, mathematics and applied technology.

problem. If a capital-scarce economy insists on equipping each employed worker with a large quantity of capital then it will not be able to equip very many workers. Secondly, and regardless of whether or not factor prices reflect opportunity costs, highly capital intensive techniques may not be the most profitable ones at the prevailing relative prices of labour and capital. If this is the case, and profits are reduced, potential savings will be lost and the rate of growth reduced. Thus if the techniques of production are excessively capital intensive both employment and the rate of growth will be reduced.

Acquired Comparative Advantage

The orothdox theory of comparative advantage has only descriptive powers. It can tell us why the structure of foreign trade of a given country is the way it is, but it cannot tell us what the country ought to export and import in order to increase its dynamic efficiency or rate of growth. The reason why the theory has so little prescriptive (or even predictive) power is that market prices frequently and sharply diverge from long-run social costs. This is particularly true in fragmented economies of the type we are discussing.[11] There seems to be a general presumption in most trade theory that comparative costs are not susceptible to change in the course of time, i.e. that a country's present comparative costs reflect natural advantages and are permanent.

There is, of course, a grain of truth in this presumption. Venezuela and Chile export petroleum and copper, respectively, because they happened to be endowed by nature with an abundance of these natural resources. They are unlikely to lose their comparative advantage in the export of these products unless their mineral deposits become exhausted—which, in Venezuela's case, is not impossible. This does not necessarily imply, however, that the Spanish American economies have or will continue to have a comparative disadvantage in the production of manufactured goods. It is quite possible that current market prices sufficiently distort the underlying social cost conditions so that the disadvantage is more apparent than real. It is even more likely that current market prices fail to reflect *potential* costs of production.

This argument is now widely accepted, although its application to the field of development policy has been mostly limited to a justifica-

[11] For a lengthy discussion of price distortions and their effect on international trade see K. B. Griffin and R. Ffrench-Davies, '*Comercio Internacional y Politicas de Desarrollo Economico*', *Cuadernos de Economia*, May–August 1964, pp. 100–118.

tion of (frequently indiscriminate) tariff protection for so-called infant industries. Spanish America, in particular, has erected a fairly large number of small-scale industries behind tariff barriers. The implications of the above argument are much more general, however. The distortions and malfunctioning of the price system suggest not only that certain industries producing for the domestic market should be protected but also that selected export industries should be subsidized. The development of new export industries may be far more important in the long run than a proliferation of highly protected and inefficient import-substituting industries.

In Chapter VI we argue that it is likely to be extremely difficult for Spanish American countries to export manufactured products to the industrialized nations of the West. The best solution may be regional co-ordination of industrial investment and the erection of common tariffs against third nations. Regardless of whether or not Latin American integration progresses it will become essential for the individual Spanish American countries to determine the 'mix' of export subsidies and tariff protection in a more rational manner. One cannot rely either on *ad hoc* interventions or on free trade because once savings become embodied in capital equipment the structure of production becomes fixed and inflexible.

In a world in which technical progress is rapidly advancing transformation curves and cost ratios can undergo sudden and substantial changes. Thus one's potential comparative advantage can alter rather quickly, and, consequently, there will be problems of continual adjustment as the economy strives to attain a new 'equilibrium'. In order for the free trade equilibrium model to give realistic results under these conditions it must be assumed that capital is homogeneous, i.e. that fixed capital can be used in the production of an indefinite number of unrelated goods. Capital, however, is durable and heterogeneous and thus the range of feasible outputs in future periods is limited by the composition of the capital stock in the present period. Orthodox theory assumes the problem away under the homogeneity postulate; movement to the new equilibrium is instantaneous. But the real problem remains: the economy cannot instantly and without cost adjust to a new production pattern. The only way to surmount this obstacle to an efficient allocation of resources is to increase economic foresight, in other words, to plan for the desired structure of production. There is no reason at all to assume that current private money costs correspond to long-run real social costs; these long-run costs will depend in part upon the development programme that is pursued. Thus it may be quite possible for Spanish America to acquire a comparative advantage in the production of manufactured goods, but she is unlikely to do so unless policies are co-ordinated to achieve that end.

3 MOVEMENTS OF EXPORT PRICES

Our preoccupation with the efficiency of the price system in Spanish America and with the high dependence of exports on only a few commodities and a few markets would lose much of its force if, in fact, fluctuations in export prices were slight and there were no adverse trends in the terms of trade. In the remainder of this chapter we will try to show that our preoccupation is genuine: movements in world prices have seriously affected most of the economies of the region.

Fluctuations of Commodity Prices

Fluctuations in the prices of primary products are a world-wide phenomenon and are not peculiar to Spanish America. A United Nations study of the instability of 39 primary commodities indicated that the average fluctuation of *unit export values* during the period 1950–61 was 11 per cent; export *proceeds* fluctuated an average of 14 per cent per annum. Additional details about ten products which play an important role in Spanish America's exports are provided in the table below.

We have already shown that most of our countries are highly

Table II:5

PERCENTAGE AVERAGE FLUCTUATION OF EXPORTS

	export unit value	export volume	export proceeds
cocoa	20	10	15
wool	16	11	15
copper metal	13	8	16
cotton	11	10	14
tin metal	10	8	10
coffee	9	7	8
sugar	7	7	9
wheat	5	3	4
bananas	3	5	5
crude petroleum	3	10	9

Source: UN, *World Economic Survey*, 1963.

specialized in the export of one or two commodities. This means that short-term movements in their terms of trade are heavily influenced by the price behaviour of a very few products. Thus we find, for example, that during the decade of the 1950s, Uruguay's terms of trade fluctuated on average 13 per cent per year, Argentina's and Chile's fluctuated 11 per cent, and in Peru and Colombia the average annual fluctuation was 9 per cent.

In general, those countries which export industrial inputs are particularly susceptible to sharp fluctuations in demand. This is because 'built-in stabilizers' in the wealthy nations have ensured that disposable income, and hence the demand for foodstuffs, fluctuates much less than GNP. Manufacturing production, on the other hand, fluctuates considerably more than GNP. Hence imports of industrial inputs such as wool and copper, which reflect changes in industrial output and not simply GNP, will fluctuate more than the latter. The volatility of demand will further increase to the extent that imports of industrial inputs are only marginal to local production, i.e. to the extent that they supplement rather than replace the output of local raw material producers.

The Declining Share of World Trade

A second general feature of international trade is the declining share of exports from underdeveloped countries to the industrial nations of the West. During the period 1953–60 the share of the underdeveloped countries in total world trade declined by over 22 per cent.[12] This tendency, as can be seen in Table II:6 below, has continued in the present decade.

The position in Spanish America, however, is even worse than the general trend suggests. The region is part of Latin America and the share of the latter in both world trade and in the trade of the underdeveloped countries has declined considerably since the end of the Second World War, i.e. Latin America's trade has grown less rapidly than the trade of the underdeveloped countries as a whole. The exports of Spanish America, moreover, appear to be growing slightly less rapidly than the exports of Latin America. That is, Spanish America's trade is a declining proportion of a falling share; the region, thus, is doubly damned.

Spanish America is playing an ever smaller role in international commerce and the region has failed to participate fully in the recent rapid expansion of world trade. Yet accelerated development almost

[12] GATT, *International Trade 1963*, 1964, p. 3. Also see A. Maizels, *Industrial Growth and World Trade*.

inevitably requires a rapid growth of imports. These imports can only be financed through a rapid increase of exports—or through substantial capital imports. Failing the latter, a relatively slow growth of exports may lead to a severe foreign exchange bottleneck and retarded development. This has become a major problem in several countries.

Table II:6

TRADE SHARES, 1960–66
(millions US dollars and percentage)

	1960	*1963*	*1966*
Exports from—			
Spanish America	5,246	5,940	7,002
Latin America	7,950	9,190	11,040
Underdeveloped countries	26,900	31,200	38,400
Total World Exports	113,400	136,000	181,400
Exports from Spanish America as per cent of exports from—			
Latin America	66·0	59·5	63·5
Underdeveloped countries	19·5	19·0	18·2
World Trade	4·6	4·4	3·9
Exports from Latin America as per cent of exports from—			
Underdeveloped countries	29·6	29·4	28·8
World Trade	7·0	6·8	6·1
Exports from underdeveloped countries as per cent of world exports	23·6	23·0	21·2

Source: IMF, *International Financial Statistics*, May 1968.

In Colombia, for instance, export receipts increased less rapidly than the population and, indeed, exports *per capita* declined by about 1·8 per cent a year during the period 1958–1966. In an attempt to ease the balance of payments constraint on growth imported consumer goods were replaced by domestic production, as a consequence of which consumer goods now account for only five per cent of total imports—the rest being raw materials, spare parts and capital goods. The capital goods sector of Colombia is still relatively small and hence a high rate of investment in plant and machinery is dependent upon the availability of imports. Any decline in the capacity to import leads immediately to a fall in investment[13] and ultimately in

[13] See K. B. Griffin, 'Coffee and the Economic Development of Colombia', *Bulletin* of the Oxford University Institute of Economics and Statistics, 1968, Part C.

the rate of growth. In other words, until Colombia is able to develop a substantial capital goods sector—either alone or as a member of a larger regional group—the level of investment and rate of growth are likely to be adversely affected by the slow growth of exports and her declining share of world trade.

The Terms of Trade

A third feature of the international trade of some underdeveloped countries is the general deterioration of their export prices and terms of trade since the end of the Second World War. In particular, most Spanish American countries experienced a very sharp fall in their terms of trade in the period 1950–1964. This deterioration was due not so much to a rise in the price of imports as to a pronounced decline in the price of exports following the end of the Korean War. Prices of manufactured imports, in fact, rose only moderately during this period.

Naturally, certain commodities fared worse than others. Between 1954 and 1962, for instance, the price of cocoa fell over 61 per cent; coffee prices declined by nearly a half; the prices of Paraguay's *quebracho* fell by 35·5 per cent; wool and cotton prices fell by a quarter and the decline in lead prices was nearly as great; the prices of wheat and of nitrates declined by over 10 per cent. The prices of bananas, beef, petroleum, copper and zinc showed no downward trend and the price of tin actually rose by over 26 per cent.

Countries which rely heavily on exports whose prices are declining are bound to suffer a decline in their export price index and their terms of trade. Considering again the period 1950–64, it can be seen, in Table II:7 that Argentina, Colombia, Ecuador, Paraguay, Peru, Uruguay and Venezuela all experienced a worsening of their terms of trade; only Bolivia and Chile—two mineral exporters—were able to improve their position. About 1962, however, the terms of trade began to improve for most Spanish American countries; the one exception was Venezuela. This can be seen in the second column of Table II:7, where changes for the period 1958–65 are indicated.

Of all the underdeveloped regions the Spanish Americans and their colleagues in the rest of Latin America complained most bitterly in the post-Korean War period about the deterioration of their terms of trade. Some commentators have tended to dismiss this as an idiosyncrasy of the region, but the facts speak for themselves. During 1954–62 the terms of trade of the underdeveloped countries as a whole declined by 13 per cent. Asia's terms of trade, however, improved by nearly 4 per cent. Thus the brunt of the deterioration of commodity prices fell on Africa and Latin America. Africa's terms of trade declined by 16 per cent and Latin America's by 25 per cent. If we exclude petroleum from the last calculation, however, we find that the decline of Latin America's terms of trade was 29·5 per cent.

Table II:7

TERMS OF TRADE

	1964 (1950=100)	*1965 (1958=100)*
Argentina	92	121
Bolivia	181	189
Chile	112	140
Colombia	77	100
Ecuador	67	88
Paraguay	86	n.a.
Peru	79	115
Uruguay	88	140
Venezuela	65	65

Sources: ECLA and UN, *Statistical Bulletin for Latin America*, Vol. IV, No. 1.

This in itself was quite serious and it may become so again if the war in Vietnam were to end and if this were to lead to a fall in the demand for the prices of several important primary commodities. Even more disturbing, however, is the fact that the volume of exports from several Spanish American countries failed to increase rapidly *despite the strong incentives provided by the extraordinary recovery in the terms of trade during the seven years of 1958–65*. The three countries whose terms of trade most improved in this period were Bolivia, Uruguay and Chile, yet the quantity of exports from these countries only increased at an annual rate of 1·0, 2·0 and 3·9 per cent, respectively. In the case of Bolivia, the volume of exports did not increase even as fast as the population, so that *per capita* export volume declined. This suggests that some Spanish American economies may have a structural export problem: they are neither able to alter the composition of exports rapidly in response to changing relative prices, nor are they able to increase the supply of traditional exports rapidly when, by chance, the terms of trade move in their favour.

In only two countries—Peru and Paraguay—did the volume of exports grow relatively rapidly. In the remainder the rate of growth was less, and usually substantially less, than five per cent a year. In the present decade, as we have shown, Spanish America benefited from rising export prices and improved terms of trade. As a result, just over half of the countries in the region were able to achieve a rate of growth of the *value* of exports in excess of five per cent a year; Venezuela, Colombia, Uruguay and Argentina, however, were not. Because of the heavy weight of these four countries the value of exports from Spanish America as a whole increased only 3·7 per cent a year during the period 1960–67.

Table II:8

VOLUME OF EXPORTS, 1958–65
(percentage annual rate of growth)

Argentina	4·5
Bolivia	1·0
Chile	3·9
Colombia	3·1
Ecuador	4·8
Paraguay	8·3
Peru	9·2
Uruguay	2·0
Venezuela	3·9

Source: UN, *Statistical Bulletin for Latin America,* Vol. IV, No. 1.

This rate of increase of foreign exchange earnings is woefully inadequate to sustain even a moderate rate of growth of national income. In most Spanish American countries the income elasticity of demand for imports is greater than unity. This means that as national income increases, the demand for imports will increase proportionately faster, and in the absence of substantial capital inflows, this demand will have to be satisfied by rising exports. In other words, the growth of exports must be more rapid than the growth of national income, in order to satisfy the demand for imported goods. Assuming, for example, that the income elasticity of demand for imports is 1·2 and the population is increasing at an annual rate of 2·6 per cent, a target rate of growth of national income of 5·0 per cent a year implies that imports will (and hence exports must) grow at a rate of 5·5 per cent a year. That is, under these assumptions, to achieve its target rate of growth of income Spanish America must increase the rate of growth of the value of exports by nearly 49 per cent. Conversely, a yearly rate of growth of exports of 3·7 per cent is compatible with a *per capita* growth rate of national income of only 0·92 per cent.

Theoretical Causes of a Decline in Export Prices

For any particular Spanish American country the cause and extent of a decline in the terms of trade will depend upon three factors: (i) the nature of the products which she produces and exports, (ii) the

Table II:9

AVERAGE ANNUAL RATE OF GROWTH OF EXPORTS, 1960–67
(percentage increase)

Argentina	4·7
Bolivia	16·3
Chile	9·1
Colombia	1·1
Ecuador	5·5
Paraguay	8·9
Peru	8·3
Uruguay	2·6
Venezuela	0·8
Spanish America	3·7

Source: UN, *Economic Survey of Latin America, 1967* Part I, (E/CN. 12/808), May 16, 1968.

degree of structural rigidity in the economy, and (iii) the bias of technical progress.

The first factor places considerable weight on a statistical finding, known as Engel's Law, that a decreasing proportion of total income is spent on food as households' incomes rise. This observation of the behaviour of individual consumers is then extended to the field of international trade where it serves as the basis for the prediction that the marginal propensity to import and the income elasticity of demand will be greater for manufactured goods than for agricultural commodities. Since the Spanish American countries import the former and tend to export the latter a structural disequilibrium in their balance of payments may be built-in to the economy. That is, given identical rates of growth of national income in two groups of countries—one of which exports manufactures and the other agricultural products—the world demand for manufactures will tend to grow faster than the world demand for agricultural products. This means that countries which export the latter will tend to generate balance of payments deficits; these deficits may tend to be accompanied by a secular decline in the price of their exports and an increase in the price of imports.

This tendency toward a secular decline in the terms of trade of countries whose exports face low income elasticities of demand could be neutralized if the increase in supply of primary products is small relative to the increase in supply of manufactured goods. In other words, given the composition of exports, a decline in their prices can be avoided only if supply increases less rapidly than world demand. As the demand for primary products grows less rapidly than world output, a decline in the terms of trade of primary producers can be avoided if, as a group, they grow less rapidly than the industrialized nations. That is, slow growth of export volume (and hence possibly of national income) may be a substitute for falling terms of trade.

Such a policy, however, is not practicable except in cases where world supplies of the primary product are located in only a few countries or are controlled by only a few international firms. Petroleum and mineral producers are perhaps the best example of this. Furthermore it is doubtful whether a policy of restraining growth to prevent a deterioration of the terms of trade, even if practicable, would be desirable as it might lead to the presence of unutilized resources and an increase in international income inequalities. It is no consolation to the underdeveloped countries to tell them they can avoid a decline in their export prices if only they would agree to remain poor![14]

The argument we have developed so far depends upon the assumption that the composition of exports is given, i.e. that the production possibilities of the economy are severely limited and the composition of output is unable to respond to changes in relative prices. These conditions were represented in Figure II:1 above by a single point transformation curve.

[14] In principle it might be desirable to co-ordinate internationally the production of primary products, at least to the extent of exchanging information about investment plans. It is unlikely, however, that the mere exchange of information would lead to a restriction of output and a rise in the price of primary products. A major reason for this—as the various attempts to negotiate commodity agreements have shown—is that the interests of the different producing countries conflict, i.e. there is no harmony of interest among the underdeveloped countries. Low cost producers are interested in expanding output and increasing their share of the world market—even at the risk of reducing prices. High cost producers, on the other hand, are interested in restraining the growth of output and maintaining relatively high prices.

Orthodox neo-classical theory would lead one to believe that a change in relative prices would induce an economy to alter the composition of production in such a way that specialization in the declining price industries would be reduced and output in the rising price—and hence relatively more profitable—industries would increase. This adjustment would help to counteract the tendency towards a fall in the terms of trade. Thus a flexible economy with a diversity of natural resources could nullify in part any tendency there might be for its terms of trade to deteriorate.

The economies we are considering, however, are characterized by a lack of flexibility, the presence of a large number of structural rigidities, and a consequent inability to shift resources from declining to expanding industries. If an economy has a transformation problem then its terms of trade will be quite arbitrary; if it is unable to react to market stimuli then, in the absence of any other adjusting mechanism, it will only be fortuitous if the country enjoys favourable terms of trade. It might 'just happen' that a country exports a commodity that is relatively high priced, but if market conditions change and export prices fall, an inflexible economy will have no alternative to suffering a decline in its terms of trade. Thus it is the combination of structural rigidity and low income elasticity of demand for exports which is deadly. Countries in this unfortunate position have very poor prospects indeed.

Structural rigidities sometimes take the form of asymmetrical reactions to price changes. A rise in price may lead to an increase in investment and an eventual expansion of output, whereas a fall in price will not necessarily lead to a reduction of output. In other words, long-run world supply curves may be kinked: supply may be elastic for favourable shifts in demand and inelastic for unfavourable shifts in demand. If this is so, then fluctuations of demand will lead to a gradual decline in the average price of the commodity concerned. This argument is illustrated in the figure below.

Assume, for example, that the price of coffee is initially p_0, as determined by the intersection of the supply and demand curves—S_0S and D_0 respectively. Assume, further, that there is an increase in demand to D_1 caused, say, by the outbreak of a war in Asia. The price of coffee will at first rise very sharply due to the fixed capacity of the existing coffee plantations. These extraordinarily high prices will encourage people to invest in this sector; output will eventually increase substantially, and the price will settle down at p_1.

The increase in supply may come from two sources. First, resources which were formerly unemployed may be brought into use in the established producing countries. For instance, a sharp rise in coffee prices might induce Colombia to combine unemployed labour with uncultivated land and thereby greatly increase her output. This sort of reaction from a major established producer, such as Brazil, could easily swamp the original price increase. Secondly, and perhaps even more important, the initial price increase might encourage foreign

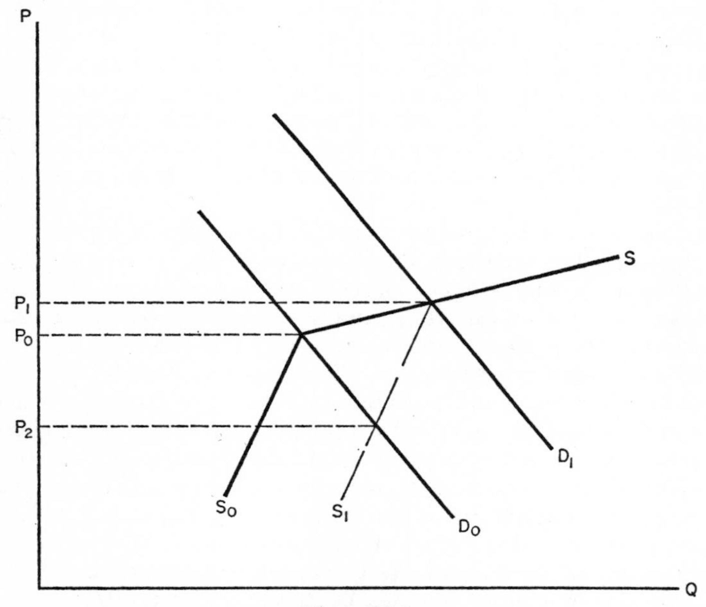

Figure II:2

producers to use *their* unemployed resources and enter the market *for the first time*. Thus a marked rise in prices may not only encourage old, established producers to expand output, it may also stimulate new competitors to enter the market. This, in fact, happened after the Second World War when Africa first became a major competitor of Latin America. The net result of this expansion may be to eliminate entirely (or more than eliminate) the price increase which provided the initial incentive to expand production. This reaction is especially likely, as in coffee, when the gestation period of investment is long, i.e. when there is a long lag between planting the trees and harvesting the first crop.

If, now, demand falls, say, to its original level, it will move along a new supply curve, S_1S. Price will not return to its initial level but, on the contrary, will fall considerably below this, to p_2. Once capacity has been installed in new plantations with long lives, variable costs of production may be very low and, hence, supply may be quite insensitive to price reductions. Once resources have become fixed in specific and long-lived capital—in this case, coffee trees—the economy becomes inflexible and is unable to transform resources from declining-price to rising-price sectors. If, in addition, the reaction to price changes is asymmetrical so that a rise in demand leads to a small increase in price and a large increase in output and a fall in demand leads to a small decrease in output and a large decrease in price, then primary producers in this situation are virtually certain to experience deteriorating terms of trade in the long run.

A third factor which helps to explain the decline in export prices of some Spanish American countries during the fifteen years after the Second World War is biased technical progress. Technical advances do not appear to be entirely random but, rather, they appear to reflect major price movements, i.e. relative factor and commodity scarcities. At the same time, technological developments are both a symptom and a cause of economic dynamism. As such they are inextricably associated with growing industrial societies. This implies that technical advances are a response to and a solution for some of the economic problems of the wealthy economies—where the scientific research and development is undertaken. There is no reason to assume that technical advances which contribute to the development of the industrialized nations will necessarily help the underdeveloped nations. On the contrary, it is plausible that the opposite will be the case.

Continuous increases in technical knowledge in the industrialized countries enable capital to be substituted for labour and primary materials—the factor of production and the commodities which the less developed countries possess in abundance. This is part of the bias or non-random development of technical progress. As is generally recognized, these increases in knowledge (i) enable the industrial economies to reduce the primary content of final products, such as the reduction of the tin content of tin cans, (ii) enable the wealthy nations to produce high quality finished products from less valuable or lower quality primary products, such as a reduction in the quality of coffee used to produce instant coffee, and (iii) enable the advanced economies to produce entirely new manufactured products which are substitutes for existing primary commodities, such as the development from plastics of a synthetic leather which 'breathes'.

These technological developments are irreversible and asymmetrical. This is what Professor Nurkse meant when he said 'crude raw materials may . . . face a kinked demand curve.'[15] The demand for many primary goods may be inelastic for price decreases but in the long-run it may be very elastic for price increases. A rise in the price of a raw material provides an incentive to undertake industrial research into ways of economizing on the commodity or substituting something else for it or producing it in the importing country. A price decline appears to provide no such incentive for investigating new ways of using the commodity. For example, a fall in the world price of copper may lead to a very small increase in the quantity demanded, but a rise in the price of copper may result in its permanent substitution by aluminium. Under these circumstances there will be a long-run tendency for the price of unprocessed raw materials to fall.

Of course, rapid increases in technological progress contribute to the general flexibility or adaptability of the innovating economy. One of the factors which explains the rigidity of the underdeveloped economies is their slow rate of technical improvement. Thus all our partial explanations of Spanish America's terms of trade and export difficulties ultimately are based upon the inability of these economies to quickly and easily transform resources. As indicated above, this so-called transformation problem is due to the specificity of resources, the low investment ratios, untrained labour and inadequate expenditure on scientific and technical education. These factors both influence and, to some extent, are reinforced by the structure of trade. A country, e.g. Venezuela in the 1950s, may enjoy good fortune; its volume of exports, terms of trade and rate of growth may be quite satisfactory. For the lucky few the transformation problem may not even be apparent. For the others, however, the transformation problem may first appear as one aspect of a balance of payments deficit; in the intermediate run it may appear as a terms of trade and export problem, and in the long-run it will appear as a growth problem. Thus one should not expect to witness a reversal of the unfavourable trade prospects of the Spanish American countries until the economic, sociological and institutional framework has been

[15] R. Nurkse, 'International Trade Theory and Development Policy', in H. S. Ellis, ed., *Economic Development for Latin America*, p. 254.

changed so as to reduce the rigidity of these economies. The foreign trade problems of the region are due in part to external conditions, e.g. tariffs and quotas in the industrial nations, but a large part is due to—and is a reflection of—the general underdevelopment of Spanish America.

CHAPTER III

Capital Imports and National Development

Ever since the fourth lecture of Sir Roy Harrod's *Towards a Dynamic Economics* economists have been discussing the relationship of long-term capital movements and economic growth in terms of simplified mathematical models.[1] It is becoming increasingly apparent, however, that both the nature of growth envisaged in these models and the assumptions underlying these models are highly objectionable. Indeed, in Spanish America the use of such models can be very misleading. The problem of these countries is not so much 'growth', i.e. expansion of a given socio-economic system, as it is 'development', i.e. rapid and fundamental politico-socio-economic transformation. However interesting 'growth theory' may be as an abstract intellectual exercise, it is very doubtful that it can be applied in Spanish America without first introducing substantial amendments.

Moreover, when one is trying to apply theory in a specific context several further considerations arise. First, the economist working in the underdeveloped countries must distinguish between foreign capital and foreign capitalists. Next, he must distinguish between the effects of a general inflow of capital and the effects on the economy of capital embodied in specific sectors and industries. Finally, he must distinguish between the effects of capital imports on the level and rate of growth of income and the subsequent effects of profit and interest repatriation on the balance of payments. Thus it is quite possible that in some circumstances general capital imports may be desirable while direct private foreign investment may be undesirable. A country may be anxious to receive foreign capital; at the same time foreign entrepreneurs may be feared for the effects they might

[1] See, for example, H. Johnson, 'Equilibrium growth in an International Economy', *The Canadian Journal of Economics and Political Science*, November 1953; reprinted as Chapter V of *International Trade and Economic Growth*.

have upon a country's capacity for change and transformation and upon the development of an indigenous entrepreneurial class.

In this chapter we propose to reconsider the role of foreign capitalists and foreign capital in the light both of theoretical reflection and the actual experience of some Spanish American countries. The traditional theory is first summarized and then modified by introducing a series of additional elements.

1 FOREIGN CAPITAL AND FOREIGN CAPITALISTS

It is usually argued that capital imports will accelerate a country's rate of growth. In the absence of foreign aid or private foreign investment a country's rate of growth (q) is determined by the proportion of its income that is saved (s) and the effectiveness of investment, or the incremental output-capital ratio (e): $q = se$. Foreign capital is assumed to supplement domestic savings and leave unchanged the effectiveness of investment, so that the growth rate rises by the amount 'fe', where 'f' is net capital imports expressed as a proportion of national income. The rate of growth of national income then becomes $q = se + fe$.[2]

Foreign capital is viewed as an addition to the physical resources of a developing economy and it is assumed that all of these additional resources are saved and invested. Thus, in the absence of capital imports the growth rate would equal only 'se', while foreign assistance permits the growth rate to rise to $(s + f)e$. Some authors go even further and argue that not only do capital imports raise the rate of investment by the full amount of the foreign assistance, they also lead to a higher rate of domestic savings (s), since the marginal propensity to save is assumed to be higher than the average.[3]

In addition to raising the aggregate investment ratio, other economists have argued that foreign capital may also play an important role in transferring knowledge from the industrial to the primary producing countries. By serving as a transfer vehicle for technical knowledge and organizational ability, capital imports may help to raise the effectiveness of investment (e). Just how strong may be the interdependence between 'f' and 'e' is open to considerable debate. First of all, the knowledge imported may not be appropriate for the

[2] A model of this type has been constructed by R. J. Ball. See 'Capital Imports and Economic Development: Paradoxy or Orthodoxy?', *Kyklos*, Vol. XV, Fasc. 3, 1962.

[3] See, for example, H. B. Chenery and A. M. Strout, 'Foreign Assistance and Economic Development', *American Economic Review*, September 1966.

economic conditions prevailing in the receiving country. Production techniques developed in capital abundant countries are neither easy to apply nor necessarily desirable in Spanish America. Their implementation is likely to result only in more unemployment and a lower level and rate of growth of income. Secondly, there are few reasons why the great amount of knowledge that is useful cannot be imported separately. Most information can be transferred through the normal trade in goods or through the movement of persons: patent rights and technical journals can be purchased; technicians can be imported on a temporary basis. Finally, capital imports, particularly foreign aid, are likely to alter the composition of investment in favour of large, lumpy, capital-intensive projects with long gestation periods, e.g. roads, dams, university buildings. This change in the composition of investment, far from raising the incremental output-capital

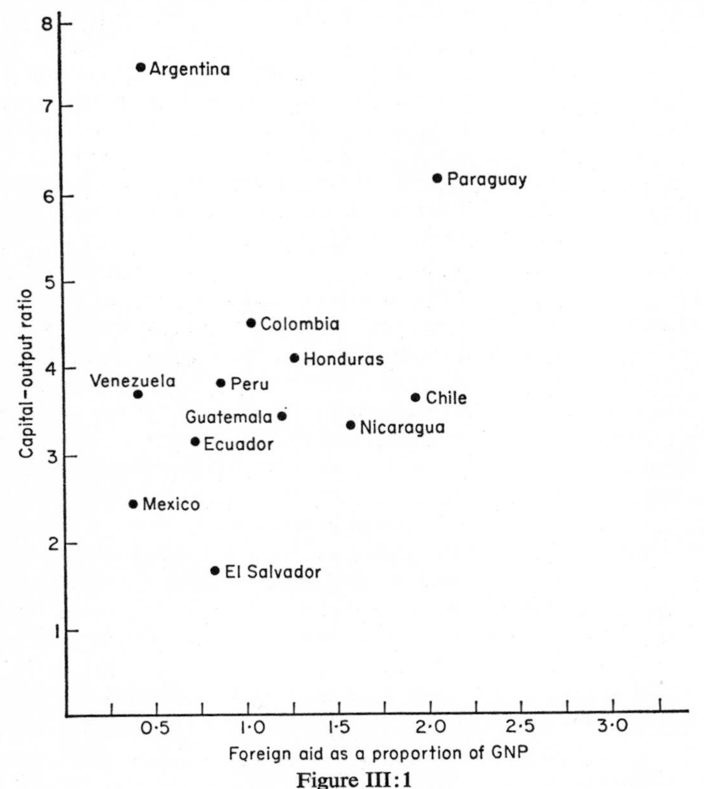

Figure III:1

ratio, is likely to lower it—and thereby reduce the rate of growth. The scatter diagram above of twelve Latin American countries (Argentina, Chile, Colombia, Ecuador, Paraguay, Peru, Venezuela, Mexico, El Salvador, Honduras, Guatemala and Nicaragua) indicates that the capital-output ratio varies positively with the amount of aid received (expressed as a proportion of GNP). The great exception is Argentina, which had a very high capital-output ratio and received little aid.

Precisely how foreign aid can lead to a rising capital-output ratio is well illustrated by Turkey's experience. The United States began a large aid programme in Turkey shortly after the end of the Second World War. This programme concentrated initially on providing tractors so as to 'modernize' Turkish agriculture. Between 1948 and 1957 the number of tractors in the country rose from 1,756 to 44,144. About a third of these tractors were used in central Anatolia to clear and cultivate marginal land previously used for livestock and forestry. This mechanization programme had two major effects. First, overgrazing on the remaining pasture lands was accentuated, over-cultivation of the new lands was encouraged and erosion was accelerated. Second, 'tractorization' increased the number of unemployed. It has been estimated that each tractor created unemployment for eight people, and from 1950 to 1955 about 350,000 agricultural workers were forced to become migrants.

It was about this time that the second phase of the US aid programme—a massive road construction project—began to yield results: between 1950 and 1960 investment in highways rose by about 1,350 per cent. Having created technological unemployment and having done almost nothing to raise agricultural yields, the peasants simply used the new roads to leave the rural areas. The major effect of investing in tractors and roads was to increase the number of unemployed and expand the urban slums. Very little additional output was generated; the value of the aggregate capital-output ratio, of course, increased considerably.[4]

It is quite possible that foreign aid may have had similar consequences in Spanish America although the research necessary to test this hypothesis has not yet been completed.

Thus '*f*' and '*e*' may be associated—negatively. Similarly, '*f*' and '*s*' may be inversely associated. If this is so, the conclusions of the standard theory may be incorrect because the assumptions of that theory are wrong. That is, capital imports and domestic savings

[4] See K. B. Griffin, 'Investment Allocation and Development in Turkey', Agency for International Development, September 1966, mimeo.

may not be independent of one another as the neo-Keynesian approach implies. Policies designed to increase the domestic savings ratio may be incompatible, especially in the short-run, with the continued flow of foreign capital. The interdependence may also run the other way: the continued flow of foreign capital may reduce the domestic savings ratio. Let us begin by considering this second alternative.

Capital Imports and Domestic Savings

If one plots on a scatter diagram the average rate of growth of gross national product over the years 1957–64 for the twelve Latin American countries referred to earlier, one finds that it is inversely related to the ratio of foreign aid to GNP.[5] The association is rather loose, but the general tendency is clear: the greater is the capital inflow from abroad the lower is the rate of growth of the receiving country. There is absolutely no support for the orthodox view that foreign aid accelerates the rate of growth.

Why does aid frequently retard growth? One reason, suggested

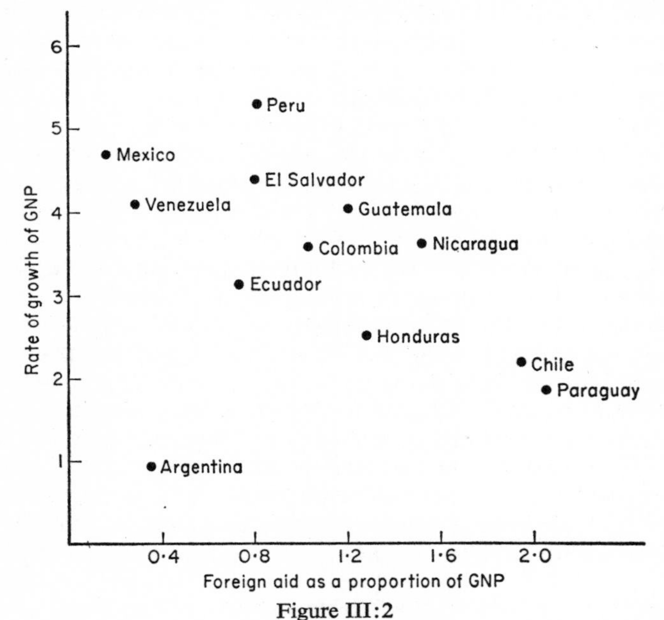

Figure III:2

[5] The diagram was first published in K. B. Griffin and J. L. Enos 'Foreign Assistance: Objectives and Consequences', *Economic Development and Cultural Change*, forthcoming.

above, is that foreign aid leads to a less desirable composition of investment and, hence, a higher capital-output ratio. Another, and more important, reason is that aid reduces the incentive to save.

The formal models usually assume that domestic savings depend upon GNP or, alternatively, upon national income *per capita* rather than upon total available resources. Thus, domestic savings are assumed either to be unaffected by capital imports or, if national income rises, to increase. In fact, one should expect the opposite to happen. A gift to an individual (say, from his uncle) or a gift to a nation (say, from Uncle Sam) is likely to be partially consumed and partially saved. That is, total investment should rise but not by the full amount of the aid. Indeed, as long as the cost of aid, e.g. the rate of interest on foreign loans, is less than '*e*', it will 'pay' a country to borrow as much as possible and substitute foreign for domestic savings. Aid may not be available in unlimited supply, of course. But given a target rate of growth in the developing country, foreign aid will permit higher consumption and domestic savings will simply be a residual, i.e. the difference between required investment and whatever amount of foreign aid is available.[6] Thus one would expect, on theoretical grounds, to find an inverse association between foreign aid and domestic savings.

Given a target rate of growth ($\bar{q}$) and an average savings ratio (s) that is insufficient to achieve the growth target, foreign assistance becomes essential. Unless, however, the marginal savings rate ($\acute{s}$) is greater than the average, a country will *never* be able to reduce its dependence on foreign aid without sacrificing the growth target. Foreign assistance can be successful in accelerating long-run growth only if it raises the marginal propensity to save. A necessary, although not sufficient, condition for ultimately achieving independence from foreign aid is that $\acute{s} > \bar{q}/e > s$.[7] Yet if our hypothesis that capital imports lead to lower domestic savings is correct, a country that relies upon foreign assistance to achieve growth may become permanently dependent and incapable of self-sustained growth.

It has become evident that foreign savings often tend to supplant rather than supplement (let alone increase) domestic savings. Speaking of Latin America, Professor Chenery notes that '. . . aid has been a substitute for savings, not an addition to investment. The savings

[6] See M. Anisur Rahman, 'The Welfare Economics of Foreign Aid', *Pakistan Development Review*, Summer 1967.

[7] See A. Sengupta, 'Foreign Capital Requirements for Economic Development', *Oxford Economic Papers*, March 1968; L. O. Geller, 'La Ayuda Extranjera: El Caso Chileno', *Desarrollo Economico*, Vol. 6, No. 24, 1967.

rate has decreased and there has been no increase in the overall rate of growth of the gross national product'.[8]

The inverse relationship between gross domestic savings and capital imports in general, i.e. aid plus private foreign investment, is very apparent in Figure III:3. This figure includes observations from fifteen Latin American nations in the period 1958–64: in addition to the nine Spanish American countries we have included evidence from six others (Mexico, Guatemala, Honduras, El Salvador, Costa Rica and Panama). As can be seen, there is a clear tendency for gross domestic savings (expressed as a proportion of gross product) to fall as foreign capital imports (again expressed as a proportion of gross

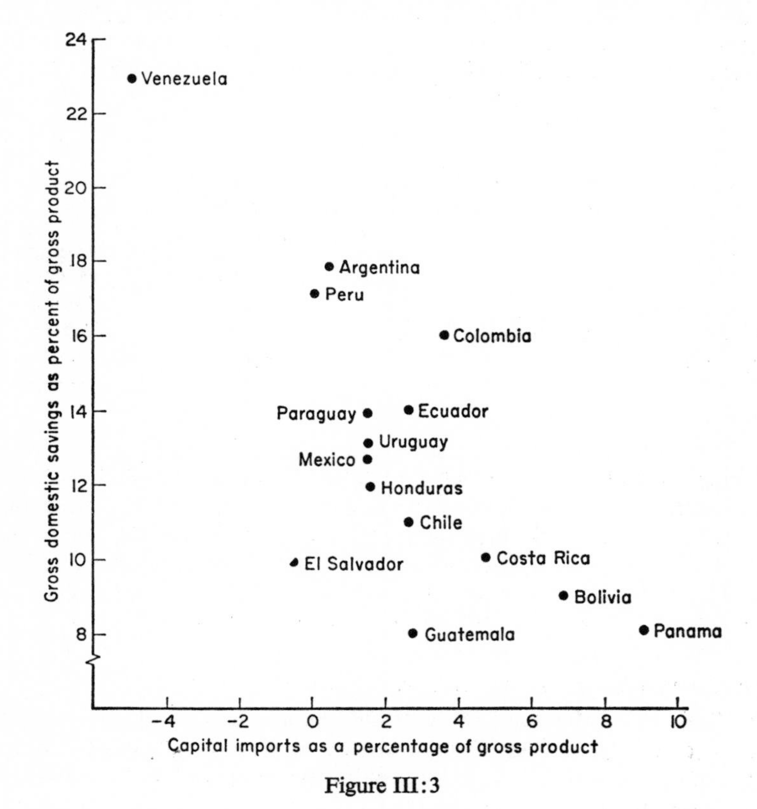

Figure III:3

[8] H. B. Chenery, 'Trade, Aid and Economic Development', in S. H. Robock and L. M. Soloman, editors, *International Development 1965*, 1966, p. 187.

product) rise. In other words, the more foreign capital a nation receives, the less it tends to save itself.

A study of a single Spanish American country, Colombia, over a period of thirteen years provides additional support for our hypothesis; there is a negative relationship between domestic savings, on the one hand, and foreign capital imports, on the other: and for every dollar of foreign aid received, domestic savings declined by about eighty-four cents.[9] The inverse relationship is very apparent if one compares the average flow of savings in the period before capital imports became important with the rate of savings in the early years of the Alliance for Progress. As can be seen in the data below, not only were capital imports associated with a proportionate decline in domestic savings, but the fall in savings was greater than the rise in aid, so that the rate of capital accumulation also declined.

Table III: 1

DOMESTIC SAVINGS AND CAPITAL IMPORTS IN COLOMBIA

	Average 1950–59	Average 1960–63
Domestic savings as per cent of GDP	22·40	16·20
Capital imports as per cent of GDP	−0·11	3·45
	22·29	19·65

Thus the available statistical evidence suggests that foreign assistance certainly does not lead to a rise in domestic savings, and probably leads to a fall. Foreign assistance in Spanish America seems to have done more to increase the region's dependence on foreigners than to accelerate economic development. Additional reasons for this tentative conclusion are provided in the sections that follow.

The Effects of Private Foreign Investment on Entrepreneurial Initiative

At least since the time of Mill economists have recognized that external capital played a major role in creating the so-called dual economies or foreign enclaves, i.e. 'outlying agricultural or manu-

[9] K. B. Griffin, 'Coffee and the Economic Development of Colombia', *Bulletin* of the Oxford University Institute of Economics and Statistics, May 1968.

facturing establishments belonging to a larger community'.[10] Many economists, however, frequently have confused income generated within a geographical region, i.e. gross domestic product, with the national product. The two are not necessarily the same, and a strategy which maximizes the former does not necessarily maximize the latter. The injection of foreign capital may increase geographical output but it does not necessarily follow that the welfare of the indigenous citizenry is thereby increased. Foreign domination of important economic sectors may introduce monopoly exploitation (in both the product and factor markets) and economic discrimination based on social or national origin.[11] Social cohesion and mobility may be destroyed. Indigenous entrepreneurs may be discouraged. Foreign investors may leave domestic entrepreneurs on the margin of economic activity; they may preempt the most profitable opportunities and retard the development of an investing class. Should the presence of investors from economically advanced societies frustrate the growth of an indigenous entrepreneurial group, the chances for long-run development will be seriously prejudiced.

The Belgian sociologist, Roger Vekemans, expresses the problem this way: foreign investments '*subrayan la carencia de capacidades nativas. Como son productos del mas alto dinamismo technologico y economico del mundo actual, recuerdan, o bien la ineptitud y atraso del ambiente local o bien las vacilantes iniciativas locales anteriores que fueron minimizadas, anuladas o derrotadas por su sola presencia*'.[12]

Professor Hagen speculates that foreign intrusion into non-Western societies has been inversely associated with technical progress and growth. As an example he cites the four largest Asian nations—Indonesia, India, China and Japan. 'Indonesia and India, which were conquered, have had the longest and most intensive contact with the West. China, which was forced to submit to the establishment of a Western beachhead along the coast and to accept trade between the interior and the West, has had the next longest and most intensive contact, while since 1600 the rulers of Japan effectively prevented almost all contact with the West. Further, most Western capital, and over the longest period, has been invested in Indonesia and India,

[10] J. H. Williams, 'The Theory of International Trade Reconsidered', *Economic Journal*, 1929, reprinted in A.E.A., *Readings in the Theory of International Trade*, p. 268 (quoting Mill).

[11] H. Myint, 'An Interpretation of Economic Backwardness', *Oxford Economic Papers*, June 1954.

[12] '*Quienes son los "Aliados para el Progresso?"*,' *Mensaje*, No. 117, March–April 1963, p. 94.

next most, over a shorter period but still a century or more, in China, but literally none in Japan until her economic modernization was well under way. But the order of entry on economic growth has been, first Japan; next seventy or eighty years later, China and India, with the growth in China . . . more vigorous than that in India; and, last of the four, Indonesia, which indeed shows no evidence of growth at present.'[13] Professor Hagen's sample, of course, is very small, and furthermore, the causal relationship between foreign investment and economic stagnation is not clearly indicated. Nonetheless, his observation is highly suggestive and increases our confidence that the relationship observed between capital imports and development in Spanish America is not fortuitous.

Private Investment and Monopoly Power: the Chilean Case

Chile has less than 13 per cent of the world copper market and about 25 per cent of world reserves. Thus Chile's share of the market in no way corresponds to her share of world copper reserves. Furthermore, the average ore content of Chile's reserves is higher than that of the United States or Canada, although not as high as that of Zambia or the Congo. Chile has an advantage over her African competitors, however, in that her transport facilities are better. These conditions have combined to make Chile an extremely low cost producer. Informed economists estimate that the cost of production in Chile is between 17 and 20 US cents a pound;[14] marginal costs probably are somewhat higher. Copper prices recently have been averaging over 40 cents a pound. Nonetheless, copper production in Chile in 1966 was only slightly higher than the levels achieved in the early 1940s. It appears that American domination of the copper industry has prevented its expansion.

Moreover, not only has total production remained relatively stagnant until very recently, production of refined copper actually declined during the 17 years after 1947. In this same period the proportion of smelted copper subsequently refined fell from 69·2 per cent to 47·4 per cent. Chile's share of world copper production declined from 19·1 per cent in 1947 to 12·7 per cent in 1966, but her share of refined copper fell by even more, indeed, by almost a half—from 10·5 per cent to 5·8 per cent.

[13] 'A Framework for Analyzing Economic and Political Change', in The Brookings Institution, *Development of the Emerging Countries*, pp. 20–21.

[14] See Chapter IV below; also see H. Castro, *Las Fluctuaciones en el Mercado de Cobre y los Ingresos Tributarios en Chile: Soluciones Alternativas*, Thesis written for the University of Chile, p. 36.

Table III: 2

CHILE'S SHARE OF THE WORLD COPPER MARKET
(per cent)

	Copper minerals	Smelted copper	Refined copper
1947	19·1	17·9	10·5
1950	14·4	13·7	9·4
1955	13·9	13·1	6·3
1960	12·6	11·8	4·5
1964	13·1	12·2	4·8
1966	12·7	n.a.	5·8

Source: ECLA, *Algunos Aspectos de la Industria del Cobre en America Latina*, May 20, 1968, p. 35.

At current prices the industry is experiencing strong competition from aluminium; some markets have been lost and many are threatened. On the other hand, the strong competition indicates that the long-run price elasticity of demand for copper must be quite high.

Thus it would appear that by substantially expanding copper production—and increasing the proportion refined—Chile could enjoy significant gains from her comparative advantage. Increased production might reduce copper prices, alleviate the pressure from aluminium, and greatly increase the quantity of copper demanded by the world market; high cost American mines might be eliminated and Chile might thereby increase her market share. Short-run instability in the copper market could be reduced by concentrating on long-run purchase agreements at fixed prices with the major buyers. Such purchase agreements perhaps could be arranged with the Socialist countries or the nationalized firms of mixed economies, and might also be arranged with the other members of the Latin American Free Trade Area, a potential market of some importance.

If the advantages of expansion are so obvious, why has production been restricted? We can mention at least three relevant factors.

First, the foreign-owned companies are afraid to lower long-run copper prices because it would eliminate the inefficient producers in the United States. The companies therefore are forced to walk a tightrope between oligopolistic pricing of copper and loss of markets from competition with aluminium.[15] Chile will continue to suffer from this subtle form of domination as long as ownership of the

[15] The Anaconda Company, one of the big producers in Chile, has hedged its risk by acquiring interests in aluminium as well.

companies and control over production decisions remain in foreign hands. Secondly, the effective tax rates in the industry were, until recently, unusually high, oscillating between 60 and 70 per cent of profits. In addition, the companies frequently have been burdened with penalty exchange rates. Thus it has not been very profitable to expand productive capacity. Third, wage rates of the miners in no way correspond to their opportunity costs. Wages and fringe benefits are estimated to be approximately four times higher in the copper mines than in other industrial activities. For obvious reasons, the copper workers are not very keen to change the management or organization of the industry. An unfortunate effect of foreign investment has been the creation of a privileged class of workers. This has reduced national cohesion and made the introduction of sweeping reforms for development more difficult. The use of exchange rate policy, high taxes and wage rates has helped to retard the expansion of the industry, but in general these measures could be justified as an attempt to ensure that some of the monopoly gains remain in the country. These largely negative and defensive measures should have been replaced long ago by positive policies for the development of the industry, but before this could be done either the Chilean mines had to be nationalized or the government had to obtain a majority stockholder interest in the subsidiary companies.

The latter solution entails a continuous outflow of profits but it will ensure continued access to technical knowledge and efficient management. Nationalization, however, could be done step by step or could be combined with a management contract and the transitional problems could thereby be reduced; featherbedding could be controlled and efficiency maintained. Ideally, the contract would be offered to the present management. Whether the American copper companies would accept such an agreement remains to be seen. Retaliation by the American government is unlikely, and at worst would involve little more than the suspension of grants and loans. Since Chile exports only 20 per cent of the copper of the Gran Mineria (and only 10 per cent of the output of the *pequeña mineria*) to the United States, the possible loss of the US market is only of marginal significance.

A third possible way of expanding the industry in Chile would be by establishing new, Chilean financed, mining enterprises. One difficulty with this approach is that capital costs are likely to be high due to economies of scale. Investments in copper would be only one of many possible uses of Chile's savings and, depending upon an evaluation of the project, conceivably might not be the best use of the given investment resources. The other alternatives avoid this problem: expansion and compensation might be self-financing out of profits.[16] In this case nationalization of the copper industry, either in whole or in part, would represent an *addition* to Chile's resources and need make no claim on the use of domestic capital.

The government of President Frei has decided to 'Chileanize' the industry. The major feature of this programme is the partial nationalization of one of the mining companies operating in the country. Specificially, the government has purchased 51 per cent of the Braden Corp., a subsidiary of Kennecott. Braden had previously stated that it would be willing to expand production by 100,000 metric tons, provided the government did not raise the profits tax above the 50 per cent level. This would have implied a reduction in Braden's profit rate from 26·9 per cent (the average over 1945–63) to 12·4 per cent. This proposal, however, was rejected by the previous government.

For purposes of compensation Braden was evaluated at $160 million, even though the book value of its assets after depreciation was only $66 million. The Chilean government not only has compensated Braden at an inflated rate, it must also contribute its share to the cost of the expansion programme. The estimated cost of the expansion programme itself suddenly and mysteriously has increased by $30 million, perhaps because Braden allowed its capital to deteriorate through lack of adequate maintenance. When all is said and done, the return to Braden's capital in the new joint venture will be about 24·1 per cent (only slightly lower than its historical average and nearly double the rate implicit in its earlier proposal) while the return on Chile's investment will be only about 8·5 per cent. Evidently, the government negotiated badly.[17]

Public Capital Imports and the Development of an Entrepreneurial Class

Some students of Spanish America might argue that it is not foreign capital so much as it is foreign capitalists that may frustrate development. If it is difficult, although not impossible, to separate the private foreign capitalist from his capital, surely—it might be argued—one can import public capital and thereby avoid the ill effects of private foreign investment. Unfortunately, however, the ill effects of capital imports cannot always be avoided in this manner. The main

[16] See below for an indication of the volume of annual profit repatriation from Chile.

[17] A detailed evaluation of the agreement is presented in the next chapter.

reason for this is that public aid is used quite consciously by the donor nations to encourage Spanish America to accept private foreign capital. Spanish America is offered a 'package' of public and private capital, and she is not allowed to choose one without the other.

The bias in favour of private enterprise sometimes is quite subtle. For instance, certain public lending institutions, such as the Export–Import Bank, may induce competition between national entities, especially non-private organizations, and foreign entrepreneurs for the same, limited, investment funds. These institutions favour private enterprise and thus private investment is encouraged at the expense of state capital formation. The composition of public and private investment usually is quite different: the former tend to concentrate on social overhead capital and, occasionally, on heavy industry, while the latter frequently engages in producing consumer goods and extracting primary materials. As in some cases the creation of an adequate infrastructure and basic industries may well have priority over the expansion of extractive industries or the production of consumer goods, the bias towards private enterprise may lead on occasion to a misallocation of investment resources.

In its eagerness to encourage private foreign initiative the World Bank goes so far as to insist that governments of underdeveloped countries open the bidding on public construction projects to foreign companies.[18] The World Bank even bends over backwards to ensure that construction projects are attractive to foreign bidders. In Chile, for example, the Bank put pressure on the government to raise the minimum bid on road building projects to five million *escudos*. The average domestic contractor, due to his limited financial resources, was unable at that time to handle a contract much larger than 200,000 escudos. Thus native entrepreneurs were squeezed out of the large state construction projects and the development of a strong entrepreneurial class was frustrated.

One of the important tasks of public policy in Spanish America is to protect and strengthen the entrepreneurial class. Under certain circumstances the importation of public capital may weaken this class. This may occur, for example, if foreign loans are used to finance investment projects which have a low capital import content, such as construction of roads, schools, houses, and hospitals. To use external loans to finance this type of internal investment the foreign exchange must be converted into domestic currency. In most cases this means the government temporarily will have to permit addi-

[18] See the statement by an IBRD economist, R. E. Carlson, in P. D. Zook, ed., *Economic Development and International Trade*, 1959.

tional imports. That is, in order to avoid inflation, tariffs, quotas and other controls will have to be reduced.[19] Additional imports, due to the sudden change in policy, will impinge upon the profitability of nationally established import-substituting industries. Native entrepreneurs will be weakened and, in societies where saving and investment decisions are interdependent, private savings are likely to decline. That is, if profits and investment opportunities decline, private savings probably will fall as well. This process clearly occurred in Chile in the period after 1956.[20] The policy of relying on foreign loans to finance investment resulted in a shift in the composition of investment from the private to the public sector, a reduction in private savings, a weakening of the entrepreneurial class and an encouragement to speculate in foreign exchange.

Attitudes to Future Capital Inflows

Having discussed the effects of past capital imports on the development of an entrepreneurial class, the level of savings, the capital-output ratio and the productive structure of the Spanish American economy, we must now consider what should be a country's policy with regard to possible future capital inflows. Two questions must be discussed: First, are future inflows desirable and, secondly, are they in fact likely to be realized? Contrary to what might be thought, it is not sufficient merely to assess the desirability of capital imports; it is also necessary, at the same time, to assess their practicability. For if it is found that capital inflows are desirable, a particular development strategy will have to be formulated to achieve that end. If, however, it is known *a priori* that the ends are unlikely to be achieved in practice, a second—different—strategy will have to be designed.

It is clear—but also trivial—that, everything else being equal, a reduction in profit leakages will ease the pressure on the balance of payments and, in this way, assist the development effort. One should not conclude from this, however, that the Spanish American countries should require foreign enterprises to reinvest all their profits in the region, for this would lead (i) to the de-nationalization of existing nationally owned industries, (ii) to a rapid rise in foreign owned enterprises relative to domestically owned firms, and (iii) to even larger profit repatriation at a later stage. Rather, Spanish America

[19] A superior alternative, or course, would be to raise domestic taxation, and consequently, government savings. For political reasons this course is seldom followed.

[20] See Institutio de Economia, *Los Efectos Economicos de los Prestamos Externos, El Caso Chileno*, Santiago, 1963.

should favour long-term low interest rate loans instead of direct private investments, loans repayable in soft currency instead of in hard, and grants instead of loans. Even if the advocacy of such a programme were completely successful, however, it seems unlikely that capital imports by themselves—even if generously available—could provide a strong impetus to development.

Increasingly, it is realized that the development process requires a strong ideological foundation—whether this be nationalism, communism or Catholic socialism. As Professor McClelland, a Harvard psychologist, has stated, 'There is no real substitute for ideological fervour.'[21] He concludes that 'ideological movements of all sorts are an important source of the emotional fervour needed to convert people to new norms. They are necessary and should be supported in whatever form is politically feasible or most congenial to the country concerned'.[22] Foreign investment, however, is not the type of activity which generates and sustains an ideological fervour; this is something which only domestic efforts can produce. Large scale participation in an economy by foreigners, in fact, is likely to frustrate the growth of an entrepreneurial class, disrupt national cohesion, and delay the appearance of a development ideology. Thus it would seem unwise to rely on capital imports as a major stimulus to economic development.

In any event, capital imports by the poor and slowly growing countries of Spanish America are not likely to be large enough to generate sustained growth. First, foreign capital appears to be attracted to the wealthier of the underdeveloped countries. A study of the period 1952–58 indicates there was a $12·9 billion flow of funds to underdeveloped countries. Of this flow, 51 per cent was private—mainly invested in the major mineral exporters.[23] On the basis of a sample of 29 underdeveloped countries, Pilvin states that 'the lower half of the underdeveloped populations received 8 per cent of the total net flow, while the lowest 70 per cent received but 15 per cent of the foreign funds. . . . The highest tenth of the underdeveloped populations alone obtained 58 per cent of total foreign funds: some three-eighths of the total flow were concentrated in the highest two per cent of the populations'.[24]

Second, foreign investments and loans seem to have been attracted

[21] *The Achieving Society*, p. 430. [22] *Ibid.*, p. 339.

[23] H. Pilvin, 'The Distribution of Long-Term Foreign Funds to Underdeveloped Countries 1952–58', *Economic Development and Cultural Change*, Vol. XI, No. 1, October 1962, p. 42.

[24] *Ibid.*, p. 45; see also H. Myint, *The Economics of the Developing Countries*, p. 180.

more often to a country *after* growth has been initiated. That is, it is becoming increasingly clear that foreign capital historically has played a very small role in assisting a country to 'take-off'. In the United Kingdom, France, Germany, Finland, Japan, etc., capital imports played virtually no part in initiating development. Even in the much publicized cases of Canada, Australia, New Zealand, and the United States foreign capital and foreign capitalists did not initiate the economic expansion but only 'joined in long after the expansion was under way and when prospects had looked bright for quite a while'.[25]

Celso Furtado mentions that 'In the Sao Paulo district, for instance, the great expansion of the railways was begun with resources of domestic origin. Foreign capital for the public services came only at a later stage, when a vigorous process of development was already under way'.[26] Thus it appears that it is the rich and growing countries that are recipients of private capital inflows. In Latin America, for example, the nine poorest countries (Bolivia, Ecuador, Paraguay, Dominican Republic, El Salvador, Guatemala, Haiti, Honduras and Nicaragua) represent approximately 13 per cent of the region's population yet received in the period 1951–60 only about 6·4 per cent of total (net) capital inflows and 4·9 per cent of direct investments.[27] In the Spanish-speaking countries of South America the four richest countries (Argentina, Venezuela, Chile and Uruguay) include 53 per cent of the area's population yet they were the recipients of 86 per cent of the region's direct foreign investments.[28]

Finally, the amount of aid the wealthy countries are willing to provide for development now seems to be declining. In 1961 the flow of assistance from the 15 DAC (Development Assistance Committee) countries was $8·73 thousand million: in 1964 it had declined to $8·66 thousand million. Meanwhile, debt servicing repayments from the underdeveloped countries, particularly from Spanish America, increased rapidly and the terms on which aid was granted, particularly the extent of tied aid,[29] became less favourable. Between

[25] K. Berrill, 'Foreign Capital and Take-Off', in W. W. Rostow, ed., *The Economics of Take-Off into Sustained Growth*, p. 293.

[26] H. S. Ellis, ed., *Economic Development for Latin America*, p. 71.

[27] UN, ECLA, *External Financing in the Economic Development of Latin America*, pp. 23, 146, April 6, 1963; (mimeo.). Also see T. W. Schultz, *El Test Economico en America Latina*, p. 25.

[28] UN, ECLA, *op. cit.*, p. 146.

[29] A study by UNCTAD of tied aid in Chile indicates that imports financed by tied aid agreements cost *at least* 12·4 per cent more than the world price (see UNCTAD document TD/7/Supp. 8, November 21, 1967, p. 10).

1962 and 1966 the official flow of financial resources from the DAC countries to the underdeveloped countries declined from 0·72 per cent of their national income to 0·57 per cent. Hence, in view of these trends, a policy of relying on foreign assistance seems neither practicable nor desirable in Spanish America.

A Development Sequence

Rather than rely on capital imports it might be easier and much cheaper in the long-run to accelerate growth by following different domestic policies which as a by-product tend to discourage foreign investments, e.g. detailed government intervention in the economy, but which increase domestic savings. It is even possible to imagine a development sequence based on such a process.[30]

Initially, in stage (1), the rate of growth of *per capita* output is negligible, the net savings rate is low—say 6 per cent—and capital imports are relatively high. At this stage foreign firms are likely to dominate important sectors of the economy—especially the foreign trade sector—and impede the emergence of an indigenous entrepreneurial class. The potential native investor is unable to compete on equal terms with his foreign counterpart; he lacks the experience and knowledge plus the tremendous financial and physical resources the latter can command. This inherent inequality frequently is reinforced by special incentives, tax preferences and investment guarantees favouring the foreign businessman. In other words, the economy is in a dependent status.

Such a situation is likely to be stable until it is disturbed by outside events (such as the importation of foreign political ideas) or the development of popular political movements. In this second stage economic and political uncertainty increase, popular leaders begin to acquire power and seriously discuss the need for 'reform'. The increased uncertainty leads to a cessation of capital imports and, quite possibly, even to a slight decline of the already low domestic savings. Stage (2) is a transition period.

In stage (3) however, the 'reforms' actually are implemented and a series of policy measures are introduced which have the effect, among other things, of sharply increasing the domestic savings ratio

[30] It must be emphasized that the 'model' which follows is not intended to predict the future; there is nothing inevitable about the sequence outlined: the 'revolution' may be a failure; foreign troops may invade the country, etc. The purpose of the model is to dramatize alternative policies facing Spanish American governments and to sketch in the possible implications of each. A diagram of the sequence is presented in K. B. Griffin and R. Ffrench-Davis, '*El Capital Extranjero y el Desarrollo*', *Revista de Economia*, (Santiago), Vol. 83–4, 1964, p. 27.

and the level of domestically financed investment. Foreign capital, perhaps stung by a nationalization programme, stays away. Stage (3) is the period of the 'take-off': the rate of growth accelerates, national entrepreneurs become a dominant class, and the structure of the economy gradually becomes transformed into a modern industrial society. At some point during this stage foreign investors may recover their confidence and re-enter the economy, perhaps investing even more than in stage (1). The form and composition of this investment, however, is likely to differ significantly from the earlier period. Less emphasis will be placed on plantation crops, extractive industries or public utilities; more emphasis will be placed on light industry, consumer services, and intermediate producer goods.[31] Foreign capital no longer will dominate the economy and foreign investors frequently will enter into partnership with national entrepreneurs. Thus in stage (4) the savings ratio has reached its new high level and foreign investment has returned to contribute to the growth process.

Looking at the sequence as a whole it is clear that over a certain crucial range there may be a conflict between policies designed to accelerate domestic savings and growth and those designed to increase foreign participation in the economy. The discontinuous movements in domestic savings and capital imports are likely to be sharper the more dramatic are the policy shifts between the pre- and post-reform periods. The extent of the 'reforms' in large part will depend upon the intensity of the desire to accelerate economic and social development. Thus the more anxious is a country to develop rapidly the less will it be able to rely on capital imports.

The Development Sequence in Mexico

Such a sequence seems to have occurred in Mexico. During the presidency of Lázaro Cárdenas, 1934–40, the long-dormant revolution of 1910 was revived by a series of reforms which increased educational opportunities, gave organized labour a stronger voice, altered land tenure institutions, and changed the political structure of the country. As a climax to these social and economic reforms the government nationalized the railroads and the petroleum industry in 1937–38. Protesting the expropriations, the 'imperialist' countries

[31] This change in the composition of foreign capital from export-based extractive industries to manufacturing industries can be highly significant. The investors in the former have nothing to gain by development and everything to lose—and hence favour the *status quo*—because it minimizes risk and uncertainty. Progress for investors in the latter, however, generally depends upon the over-all rate of growth of the economy, and hence the interests of foreign capital and national development conflict less sharply.

responded with an international boycott. Far from preventing the Mexican 'take-off', the boycott accelerated development by forcing the Mexicans to create an efficient entrepreneurial class. Between 1939 and 1950 the average ratio of net savings to Gross National Product was 11·2 per cent.

During the period 1935–38 there was a pronounced outflow of capital from Mexico.[32] The book value of US direct investments in Mexico declined sharply from $683 million in 1929 to $316 million in 1946; they rose rapidly in the next decade, however, and by 1956 they had nearly regained the previous peak.[33] Clearly the policies of the Cárdenas era did not permanently reduce the attractiveness of Mexico to foreign investors and they undoubtedly increased the speed of development. As Professor Higgins has indicated, 'Once the Mexican economy began to move and new opportunities for profitable investment appeared, foreign capital flowed into the country, accelerating the rate of economic expansion'.[34] 'Between 1939 and 1944 (foreign) investments plus reinvested earnings averaged only $10 million annually while in the years 1949–54 the annual average increased to $22 million, or about 130 per cent higher than the yearly average in the preceding period.'[35] Thus, on the basis of the available evidence, Mexico entered stage (3) in the late 1930s and began stage (4) around 1945. The lag of nearly ten years between stages (3) and (4) probably was due in large part to the intervention of World War II, although the inexperience of foreign businessmen in adjusting to social revolutions also may have been a factor.[36]

Capital Imports and the Alliance for Progress

An optimist might imagine that the stylized sequence presented above has been operating recently in Spanish America. The symptoms

[32] For a general discussion of the role of foreign investment in Mexico see A. Navarrette, '*El Crecimiento Economico de Mexico y las Inversiones Extranjeras*', *El Trimestre Economico*, 1958. For a discussion of the role of nationalism as an 'input' in development see M. Nash, 'Economic Nationalism in Mexico', in H. G. Johnson, ed., *Economic Nationalism in Old and New States*, 1968.

[33] US Department of Commerce, *U.S. Investment in the Latin American Economy*, p. 180.

[34] *Economic Development*, p. 69.

[35] IBRD, *The Economic Development of Mexico*, p. 137. The figures cited in the quotation exclude the 'characteristically erratic movements of inter-company accounts'.

[36] One of the difficulties of testing the historical existence of the suggested stylized sequence is that frequently adequate data on an economy do not appear until stage (3). It almost appears that while the politicians were organizing a social revolution, the economists were deciding to organize a National Statistical Service!

indicate, if interpreted optimistically, that the region is in stage (2): political agitation has increased, 'reforms' are being discussed but not implemented, and foreign aid has declined.

There was also a decline of private foreign investment in the early 1960s. This can be attributed to the fear and uncertainty created in the minds of businessmen by the Bolivian and Cuban revolutions. The uncertainty was increased, and the decline in investment probably accentuated, by violence in Venezuela and Colombia in the early years of the decade and by the reformist pronouncements of the Alliance for Progress. In other words, both the initial reduction of private foreign investment and the creation of the Alliance for Progress were a reaction to social revolution within the hemisphere.

The tragedy of the Alliance is that the attempt by the US government to accommodate itself to Spanish American realities reinforced the fears of foreign businessmen without substantially changing the economic prospects for the region. In its early years the Alliance loudly called for tax, educational and agrarian reforms. It pointedly refrained from mentioning, however, the need to reform the (foreign owned) raw material industries. Nonetheless, the US State Department, at least temporarily, appeared to have become somewhat less afraid of the word 'revolution'[37] and implied frequently that most Spanish American nations needed more progressive and popular governments. Unfortunately, talk of reform produced very little beneficial activity within the Spanish American republics but resulted in, or at least was accompanied by, an initial fall in private capital inflows.

It was at this point that US policy was reversed. President Johnson's Administration became much less interested in encouraging modest reforms and much more concerned with maintaining 'stability'. This shift of emphasis was made explicit by the Resolution of the US House of Representatives of September 20, 1965 which endorsed the unilateral use of force by the United States or by any other Western Hemisphere country to prevent a communist takeover. The governments of five of our Spanish American countries objected to this Resolution; Paraguay, Bolivia, Argentina and Ecuador did not. These events were accompanied by a general swing to the 'right' in Latin America—no doubt stimulated by US military intervention in the Dominican Republic and the right-wing

[37] Indeed, the American government was delighted when Chile's Christian Democrats won the last presidential election under the slogan '*Revolución con Libertad*'.

military *coup d'état* in Brazil—and this led to the revival of private capital inflows, as can be seen in Table III:3.

Table III:3

US PRIVATE DIRECT INVESTMENT IN FIVE SPANISH AMERICAN COUNTRIES
(book value in millions of dollars)

	1963	*1966*
Argentina	829	1031
Chile	768	844
Colombia	465	576
Peru	448	518
Venezuela	2808	2678
Total	5318	5647

Source: Agency for International Development, Statistics and Reports Division.

It is perhaps ironic that the privileged groups in Spanish America have approved of the Alliance because they believe it will not work, i.e. there will be no fundamental social and political changes, while the Marxists have always opposed the Alliance because they are afraid it will work, i.e. that it will perpetuate and reinforce governing minorities. Both groups, of course, have been correct.

American businessmen, opposed to 'reform', 'revolution' and 'socialism' and fearful of nationalization programmes withdrew their capital from the region and insisted, successfully, that the US should use its aid programmes to foster their views of economic organization. In the words of one recent report, the 'encouragement of private enterprise, local and foreign, must become the main thrust of the Alliance'. The United States 'should concentrate its economic aid program in countries that show the greatest inclination to adopt measures to improve the investment climate, and withhold it from others until satisfactory performance has been demonstrated'.[38] Thus it has become evident that continued capital inflows are likely to be contingent upon the maintenance of a particular form of economic organization, 'free enterprise capitalism'. This particular economic system has been singularly unsuccessful in the past in solving Spanish America's development problem. Yet these countries now may be forced to choose either foreign capital and foreign

[38] The Rockefeller Minority Opinion, part of the report from the Commerce Committee for the Alliance for Progress. The Minority Opinion was later endorsed by the Chairman of the Committee.

capitalism or economic planning based entirely on domestic savings. Neither alternative is entirely satisfactory, but if forced to choose between them there is much to be said for electing the latter.

This is particularly true in view of the fact that foreign aid has not been forthcoming in the amounts promised and, indeed, has declined slightly. The original programme for the Alliance for Progress as negotiated at Punta del Este in 1961 envisaged an inflow of capital into Latin America of $20,000 million over 10 years. Most of this was to be provided in the form of foreign aid, i.e. AID grants and loans, Food for Peace commodities, loans from the Export-Import Bank, Peace Corps volunteers and assistance from the Social Progress Fund of the Inter-American Development Bank. A mid-term evaluation of the programme, however, indicated that in the first five years of the Alliance, US government aid commitments were only $4·9 billion, and disbursements were considerably less.[39] Presumably it was for this reason that President Johnson announced in 1965 that the Alliance would be extended beyond the original ten years.

The evolution of Alliance for Progress aid in Spanish America can be seen at a glance in the table below. We have compared aid disbursements in 1962 (the year following the signing of the Charter at

Table III:4

AID DISBURSEMENTS UNDER THE ALLIANCE FOR PROGRESS
(millions, US dollars)

	1962	*1966*
Argentina	84·2	36·2
Bolivia	34·6	21·7
Chile	102·8	92·6
Colombia	64·1	91·5
Ecuador	14·9	23·8
Paraguay	4·7	6·3
Peru	29·1	52·6
Uruguay	5·1	4·6
Venezuela	62·0	48·3
TOTAL	401·5	377·6

Source: Informe de los Estados Unidos de America al Consejo Interamericano Economico y Social, April 1967, Table 6.

[39] Senator Ernest Gruening, *United States Foreign Aid in Action: A Case Study*, US Senate, Subcommittee on Foreign Aid Expenditures of the Committee on Government Operations, Washington, 1966.

Punta del Este) with disbursements five years later, in 1966. Over this period aid disbursements declined in five out of the nine countries, namely, in Argentina, Bolivia, Chile, Uruguay and Venezuela; they rose in Colombia, Ecuador, Paraguay and Peru. For the group as a whole, total disbursements fell by about six per cent. In *per capita* terms, of course, the decline was much sharper.

Even these figures, bad as they are, fail to indicate how ineffective has been the Alliance for Progress and other programmes of external assistance. In the first place, the net inflow of foreign capital is low and falling because an increasing proportion of the gross external receipts is being used to finance amortization and interest payments. In 1961, for instance, Chile received a gross inflow of $94·3 million from AID, IBRD, IDA, the X–M Bank and the Inter-American Development Bank; the net flow, however, was only $71·9 million. By 1967 the gross inflow from the same sources had risen slightly to $101·6 million, but the net inflow had declined by nearly a half to $39·4 million.[40]

Secondly, contrary to its original purpose, the Alliance no longer is using aid as an incentive to undertake structural reforms. Let me illustrate this with another example from Chile. In the first two years of the Alliance, AID used the promise of a loan of $35 million as an incentive to get the ultra-conservative régime of President Alessandri to pass a land reform law. This pressure was partially successful and resulted in the 1962 Agrarian Reform Law, although no serious attempt was made to implement the law. Subsequently, during the new reformist régime of President Frei, AID 'became silent on agrarian reform either as an indicator of Chile's self-help measures or as a condition for further US assistance'.[41]

The alternative of relying on foreign capital thus offers little hope for rapid development. The target of the Alliance for Progress was to achieve a minimum rate of growth of *per capita* income of 2·5 per cent a year. Yet in only two years since 1961, viz., 1964 and 1965, has the minimum target been achieved. Last year, 1967, the *per capita* rate of growth was only 1·5 per cent. The Alliance is obviously failing. In an attempt to prevent this, a meeting of most of the heads of state of the western hemisphere nations was held at Punta del Este, Uruguay, in 1967. At this 'summit' conference an action programme was formulated, the principal goals of which were the

[40] V. Tokman, '*El Financiamiento Externo, Su Componente de Ayuda y los Efectos Sobre la Formacion de Ahorros y la Capacidad de Pagos: El Caso Chileno*'. mimeo., August 1968, Table 2.

[41] Gruening Report, *op cit.*, pp. 102–103.

beginning of a Latin American Common Market in 1970 and its 'substantial operation' by 1985;[42] promotion of multilateral development projects; modernization of agriculture; expansion of education and medical facilities; and the 'elimination of unnecessary military expenditures'.

The results have not been encouraging. Military expenditure has increased; the impetus to regional integration has diminished; food production has continued to lag seriously behind the growth of population, particularly in Chile, Colombia and Peru. Education has not received the attention it deserves, with the notable exception of Venezuela, where nearly 13 per cent of the national budget is spent in this sector. Exports have remained stagnant: in 1967 export receipts increased only 2·7 per cent, i.e. no faster than the population. Finally, the contribution of foreign aid has been slight: in 1966 grants and loans to Latin America net of amortization and interest payments were $800 million, compared with $2·3 thousand million in Africa and $3·5 thousand million in Asia and the Middle East; debt servicing accounted for 75 per cent of the gross capital inflow.[43]

Few economists are prepared to argue that capital imports are wholly bad. Some are prepared to argue, for the reasons given in this chapter, that domestic investment is better, and policies which encourage one may discourage the other. That is, the domestic propensity to save is not 'given'; it is a variable. The standard models discussed at the beginning of the chapter usually assume the savings ratio is constant and, in so far as it varied, it varied independently of capital imports or, in a special case, rose as a result of them. Particularly in the short-run, however, policy decisions which have the effect of increasing domestic savings may have the opposite effect on capital imports. But if the policy changes are successful and, in fact, do accelerate growth, capital imports probably will not be influenced much in the long-run. One should conclude therefore that governments of Spanish America should encourage foreign capital imports only after all other policies have been determined. Capital imports at best are only a marginal element in the growth process; the main burden of development will have to be carried by native decision makers using domestic resources. Hence domestic policies should be established first and foreign capital can adjust to them; countries should not adjust their domestic policies to the dictates of foreign capital.

[42] See Chapter VI below for a discussion of regional integration.

[43] UNCTAD Secretariat, 'The Outlook for Debt Service', (TD/7/Supp. 5), p. 71.

2 FOREIGN CAPITAL AND THE BALANCE OF PAYMENTS

Conventional theory would have us believe that a country receiving external capital would enjoy a substantial import surplus on current account. Only later, when repatriated profits and interest charges on past investments exceed the new capital inflow, would an export surplus be necessary to finance these leakages. Yet the Burmese economist, H. Myint, has shown that more often than not there were repatriated profits 'when there is no evidence of a previous inflow of capital in the form of import surpluses'.[44] How can this be? Two extreme possibilities may be cited.

First, foreign control of an enterprise may be obtained with domestic capital. That is, there may never have been a capital inflow; funds may have been raised in the capital market of the host country, e.g. through domestic banks.

That is what appears to have happened to the Chilean nitrate industry in the nineteenth century. The government decided to denationalize the holdings in 1881 and the way was left open to foreign enterprise and speculation. John T. North acquired control of the industry with loans granted by the Bank of Valparaiso. This institution and 'other Chilean lenders supplied 6,000,000 pesos to North and his associates to buy up the nitrate certificates and the railroads of Tarapacá'.[45]

In cases such as the above the underdeveloped economies obtain all the disadvantages of foreign economic domination without receiving a compensating advantage of a net capital inflow. The repatriated profits, in fact, are nothing more than a type of capital flight. Foreigners provide only organizational ability; the domestic economy provides capital and all other inputs, yet the profits of the firms are sent abroad just as if they represented a return on foreign capital. Organizational ability, of course, is worth something, but it

[44] 'The Gains from International Trade and the Backward Countries', *Review of Economic Studies*, 1954–55, No. 58, p. 138. See also P. Baran, *The Political Economy of Growth*, pp. 179–200.

[45] A. Pinto, *Chile, Un Caso de Desarrollo Frustrado*, p. 55, quoting F. Encina. Translated by the author. The copper industry also is somewhat ironical. Chilean copper did not come under foreign control until around 1900. At that time the private Chilean owners sold the mines and refineries to foreign interests. This occurred, incidentally, during a period of rising copper prices associated with the world-wide expansion of the electricity industry. The Braden Copper Co. was established in 1904 with a capital of $4 million. In 1964 the Chilean Government proposed to buy back *half* of Braden for about $82 million. See M. Lazo D'Arcangeli, *La Exportacion Chileno de Cobre durante el Periodo 1810–1910*, University of Chile thesis, 1964.

is doubtful whether it is worth a continuous outflow of profits; a high salary should be sufficient.

On the basis of a questionnaire circulated by the University of Oregon, it was determined that of the seventy-two companies reporting on their 'sources of financing of foreign affiliates', in 1959, thirty-three companies replied they borrow 'all or all possible' in the host country.[46] Between 1957 and 1959, 'host country funds' represented 17·5 per cent of all sources of funds for US direct investment in Latin America. Over the same period the host country provided 25·8 per cent of the funds for all US direct investment overseas.[47] Not all of the host country funds, of course, were raised through the banking system; some were equities and thus did not entail any subsequent capital outflow. Nonetheless the presumption exists that in many cases foreigners have gained control of domestic resources—and exported the subsequent profits—without first having imported capital.

Even if the original foreign investment can be attributed to a capital import and the receiving country consequently experienced an import surplus, it does not follow necessarily that subsequent increases in foreign investment will have similar balance of payments effects. If the increased foreign investment is due to, say, reinvestment out of profits or to a change of residence of a native capitalist, foreign control of domestic resources will increase without any additional importations of capital. In such cases it is not at all clear that the foreign contribution to the 'recipient' nation is sufficiently large to justify permanent repatriation of profits.

The second possible explanation for the absence of an import surplus is that foreign investors may demand an extraordinarily high rate of return on their investments. From the point of view of the capitalists, foreign investment is an unusually risky activity which requires a high rate of amortization. Thus foreign capital may be turned over very rapidly in sectors which yield extraordinarily high profits. These high profits, in turn, may be due in large part to monopolistic manipulation of costs and prices, or they may be due to a natural monopoly, i.e. a lack of close and cheap substitutes.

The substantial leakages of interest payments, dividends, and profit repatriation—not to mention the repayments of principal—greatly reduce the impact of foreign investment on the host economy and directly accentuate the problem of the balance of payments. Further-

[46] R. F. Mikesell, ed., *U.S. Private and Government Investment Abroad*, p. 96, Table IV–6.

[47] *Ibid.*, p. 99.

more foreign firms are likely to have a propensity to import far above the national average. This, too, reduces the 'multiplier effects' of foreign capital on the productive structure of the economy and further aggravates the payments difficulties. Little impetus is likely to be given to domestic suppliers. Managers, technicians, equipment, and many other inputs are imported. In some cases, even the labour force was imported, as occurred for example with the importation of East Indians to work the sugar plantations in what was then British Guiana.[48]

We noted above that the net contribution of foreign aid to the capital resources of Spanish America is rather small. Evidence is also being accumulated, gradually, which shows that the profit leakage from foreign private investment is remarkably large. It is rare indeed that new investment in the underdeveloped countries exceeds repatriated profits. There has been a consistent net capital outflow from the low-income, capital-poor countries to the high-income, capital-abundant countries. Data from the US Department of Commerce, for example, indicate that over the half decade 1956–60 United States businessmen invested slightly over four and a half thousand million dollars in the underdeveloped countries, yet during the same period over seven thousand million dollars of profits were repatriated. Thus on private account there was an outflow of capital net of investment of almost $2·7 thousand million in five years.

In Latin America, if anywhere in the underdeveloped world, one would expect capital inflows to exceed repatriated capital. The United States has had a long and profitable association with the Spanish American Republics. North American businesses have developed extensive contacts in the region, have experience in working with Spanish American institutions and laws, and in practice have encountered remarkably little hostility to foreign investment. Expropriations have been infrequent and eventually have been followed by compensation in all cases except the Cuban. Yet here, too, we see that over the twelve year period 1950 through 1961 the region as a whole repatriated more capital than it received. This may be seen in Table III:5.[49]

The net inflow of $1829 million on public account was not nearly sufficient to compensate for the $3910 million outflow of capital on private account. The Latin American region was a source of capital

[48] See R. T. Smith, *British Guiana*, pp. 42 ff.

[49] The data in Table III:5 refer specifically to Argentina, Brazil, Chile, Colombia, Mexico, Peru and Venezuela. These countries comprise nearly 90 per cent of the population and income of the entire Latin American region.

Table III: 5

CAPITAL MOVEMENTS IN LATIN AMERICA, 1950–61

	Total 1950–61	Annual Average
(1) Net new direct US investment:	US$2965m	US$247m
(2) Profit and interest remittance on (1):	6875	573
(3) Net movement of private capital:	−3910	−326
(4) Total US aid:	3384	282
(5) Official US debt repayments:	1151	96
(6) Interest on debt to US Government:	404	34
(7) Net movement of public capital:	1829	152
(8) Net movement of all capital:	−2081	−174

Source: US Department of Commerce and Agency for International Development.

to the United States throughout the period. It is interesting to note that the Alliance for Progress expects private investors to play an important role in developing Latin America. The target of $300 million a year on private account has been set. It is not clear whether this $300 million is supposed to be a net contribution of capital, i.e. net of profit and interest remittances, or only gross. In either case the target is unrealistically high and it is very doubtful that it will be achieved. Direct investment averaged only $247 million a year in the period covered by Table III: 5, and in eight of the twelve years private direct investment was less than $200 million. In the first five years of the Alliance for Progress the inflow of new private capital into Latin America has been about $200 million a year—excluding Venezuela, where the inflow was negative.

Another way to consider the balance of payments effects of foreign capital is to express the service payments (interest, amortization and profit repatriation) on long-term capital imports (both public and private) as a percentage of export earnings (both visible and invisible). According to ECLA's data, between 1951–55 and 1961–65 service payments on foreign capital rose from 3·2 to 30·0 per cent in Argentina, from 20·0 to 44·5 per cent in Chile, from 7·5 to 29·4 per cent in Colombia, from 9·5 to 19·0 per cent in Peru, and from 4·4 to about 10 per cent in Uruguay. In other words, payments to foreign capital in Spanish America are rising considerably faster than exports.

A critic might argue that it is not legitimate to lump all the underdeveloped countries into one group, or even to look at Latin America or Spanish America as a whole. Differences between countries are important. Those countries which have abundant resources for development and which show no intention of nationalizing foreign enterprises will be net recipients of foreign capital; countries which are hostile to foreign investment will experience large-scale capital repatriation. Once again we find that this generalization probably is incorrect. Let us take Chile as an example: (a) from 1958 until September 1964 a conservative coalition was in power; (b) the country is politically stable and democratic; there is no revolutionary tradition; (c) there are tremendous resources to be developed. In addition to the copper reserve mentioned earlier, the country has great potential in exports of wine and of iron ore; she has unexploited forest resources and could become an important producer of

Table III:6

CHILE: CAPITAL FLOWS ON FOREIGN PRIVATE ACCOUNT
('000 US dollars)

Year	(1) Net Private Foreign Invest.	(2) Repatriated Profits (a)	(3) Net Private Capital Flow: (1)–(2)
1944	−3,612	19,755	−23,367
1945	−9,158	15,381	−24,539
1946	−3,039	26,025	−29,064
1947	−3,243	44,500	−47,743
1948	12,180	56,339	−44,159
1949	29,607	30,167	−560
1950	20,492	35,024	−14,532
1951	38,497	48,233	−9,736
1952	39,509	46,793	−7,284
1953	14,977	25,075	−10,098
1954	−3,946	29,972	−33,918
1955	2,534	63,753	−61,219
1956	12,892	81,807	−68,915
1957	43,019	38,817	+4,202
1958	55,362	32,530	+22,832
Total	246,071	594,169	−348,098
Average	16,405	39,611	−23,207

Source: Calculations based on data published by the Banco Central de Chile *Balanza de Pagos* 1958. (a) includes figures only for the *gran minería del cobre, gran minería del hierro* and *salitre y yodo*. The Table ends in 1958 because methodological changes at the Central Bank after that date make further comparisons difficult.

cellulose; her bountiful fishing grounds have not been fully developed. In spite of these investment opportunities Chile has suffered a large outflow of capital on foreign private account. This is shown in Table III:6.

Net investment was negative in five of the fifteen years covered by the Table. Repatriated profits exceeded new investment in thirteen of the fifteen years. Over the entire period there was a net outflow from Chile of $348 million or a yearly average outflow of $23 million. Profit leakages, however, have not necessarily frustrated development. In so far as foreign investment creates wealth and provides a flow of goods and services which *otherwise would not have been provided* capital imports are a contribution to the economy and profit repatriation *per se* is not objectionable. This surplus over costs becomes of interest mainly for its effects on the balance of payments and as an indication of the potential benefits of expropriation. But beyond a certain point we must agree with Mr W. Rosenberg[50] when he argues, 'The heavier the interest burden on foreign investment, the greater is the likelihood of a country having to borrow to meet that interest burden. Once embarked on this slippery slope, the action of compound interest starts to work, and foreign investment tends to become a growing burden which is no longer compensated by the advantage of increased availability of real resources. . . . Thus while on the surface foreign investment may stimulate a country's development in certain fields, looking at it over a period of years, excessive foreign investment leads to a reduction of growth.'

It is important, however, that the direct effects of foreign investment and foreign aid on growth and the secondary effects of profit repatriation and debt servicing on the balance of payments should be separated. This does not imply that the state of the balance of payments has no influence on growth and development. On the contrary, a persistent payments difficulty can easily frustrate economic development—particularly if the measures adopted to control the balance of payments result in a lowering of the level of income and investment. This has occurred frequently in Spanish America and is the topic of Chapter V.

Summary

The orthodox conclusion that capital imports will accelerate economic development is not always valid. Under certain circumstances

[50] 'Capital Imports and Growth—The Case of New Zealand—Foreign investment in New Zealand 1840–1958', *Economic Journal*, March 1961, pp. 106, 108.

foreign capital, whether public or private, may fragment the economy, introduce monopoly elements into the society, discourage the development of a native entrepreneurial class, lower the domestic savings ratio, raise the capital-output ratio and cause subsequent balance of payments problems. This last problem can be avoided in part if foreign enterprises reinvest a substantial proportion of their profits in the host economy, but this, in turn, only causes further difficulties, *viz.* growing foreign control of the economy and denationalization of local industry. In Central America this process already has advanced very far. Foreign investment has penetrated not only into large industries but into small and medium industries as well. This phenomenon has been associated with the acquisition by foreigners of established firms managed for many years by local businessmen. In effect, private foreign investment has converted small local entrepreneurs into rentiers and thereby retarded the development of an indigenous capitalist class.[51]

The national interest may not be compatible with a large inflow of foreign capital. The encouragement of a development ideology and the introduction of widespread social and economic reforms necessary to raise the domestic savings ratio may be extremely difficult in countries where capital imports provide a large part of the finance for growth. Thus, over a certain range, policies designed to encourage capital imports may be in conflict with policies designed to raise domestic savings and alter the *status quo*. When such a conflict occurs, the case for concentrating on domestic savings is strong, for if the policy changes are successful in raising the savings and growth rates, capital imports are unlikely to be adversely affected in the long-run.

[51] See Committee of Nine, Alliance for Progress, *Informe Sobre los Planes Nacionales de Desarrollo y el Progreso de Integracion Economica de Centro America*, Washington, 1966, pp. 32, 115–16.

CHAPTER IV

Mixed Enterprises and Foreign Investment

In the last chapter it was argued that Spanish America would be unlikely to achieve a rapid rate of development and a satisfactory distribution of output if reliance was placed on capital imports. Private foreign direct investment, in particular, was shown to have numerous negative effects, and it was suggested that the disadvantages of relying on foreign investment frequently exceed its advantages. Even among those who accept our basic argument, however, it is commonly believed that mixed enterprises combining foreign capital with either public or private local capital are able to reduce substantially the undesirable features of capital imports.

This, implicitly, was the opinion of the Christian Democratic Party in Chile when it came to power in late 1964 and proceeded immediately to 'Chileanize' the copper industry. A detailed discussion of this programme is interesting because it enables one to examine the three different policies applied by the government in its efforts to expand production: (i) the government bought a controlling interest of 51 per cent of the large mine owned by the Kennecott Corp.; (ii) the government acquired a minority interest of 25 per cent in a joint venture with the Cerro Corp. to develop a new mine; and (iii) the government used fiscal policy to encourage the Anaconda Co. to expand production at Chuquicamata, its large mine in the north. We shall analyse these three policies and compare them with a policy of nationalization as advocated by the Popular Front.

The results of Chile's experiment with mixed enterprises will undoubtedly provide useful information to governments in other Spanish American countries which might contemplate partial or complete nationalization of existing foreign enterprises (e.g. the petroleum industry in Venezuela) or allow direct foreign investment to participate in a nationally owned enterprise (e.g. the tin mines in Bolivia).

1 BACKGROUND TO 'CHILEANIZATION'

As indicated in Chapter III, copper is one of the most important economic activities in Chile. It accounts for about eight per cent of gross domestic product, sixty per cent of exports and eighty per cent of government tax receipts which accrue in the form of foreign exchange. Chile, in turn, is one of the most important suppliers of copper. Her output—over half a billion short tons—accounts for about fifteen per cent of world production and Chile's reserves represent 22–30 per cent of those presently available.

Copper exports and prices are subject to considerable variation—sales fluctuate on average 22·1 per cent per annum—and this leads to instability in government receipts, government expenditure and GDP.[1] These fluctuations have their origin both on the supply side (e.g. strikes and political disturbances) and on the demand side (i.e. fluctuations in the level of industrial production in Europe and North America). Unlike many commodities exported by the underdeveloped countries, however, the long-run prospects for copper are quite encouraging.

In the last half century changes in relative prices have greatly favoured aluminium at the expense of copper: in the period 1913–59 the price of copper rose 104 per cent while that of aluminium rose only 14 per cent.[2] Copper lost several markets to aluminium and it is feared by some people that this process of substitution will continue. These fears probably are exaggerated, however. First, it is unlikely that relative price changes between copper and aluminium will be as drastic in the future as they have been in the past. Second, recent econometric studies indicate that both the price elasticity of demand for copper and the cross elasticity of demand are very low.[3]

More important, the income elasticity of demand for copper is fairly high since demand rises with the growth of manufactured goods, although at a slower rate. The growth rate during the period 1947–66 was 4·6 per cent per year and there is no reason to believe that this rate will decline in the next thirty years. Indeed, as long as

[1] See A. I. MacBean, *Export Instability and Economic Development*, 1966, ch. 8 and C. W. Reynolds, 'Domestic Consequences of Export Instability', *American Economic Review*, May 1963.

[2] A. Maizels, *Industrial Growth and World Trade*, 1963, Ch. 9.

[3] J. Correa H., '*Determinación Estadística de la Demanda Mundial de Cobre*', VIII Meeting of Technicians from Latin American Central Banks, Buenos Aires, November 1966, mimeo.; T. C. Thomas, *Variation in Copper Usage in the United States*, Ph.D. thesis, M.I.T., 1964.

world industrial output is maintained at a high level the demand for copper will be high.[4]

In addition to the rapid growth of copper demand, the source of copper supply is changing rapidly in favour of the non-industrial nations. Between 1929 and 1959 the output of copper in the industrial nations remained unchanged, although it still accounted for over half the world's supply at the end of the period. Production in the rest of the world, however, and especially in Central Africa, rose by 378 per cent.[5] The reason for this shift in the sources of supply is, of course, the lower costs of production in the underdeveloped countries.[6]

In the post-war period 1947–66 the share of Western Europe and the United States in world copper production declined from 38·7 per cent to 27·6 per cent. The share of the Soviet zone countries more than doubled from 7·9 to 16·3 per cent. The share of the underdeveloped countries rose only slightly, from 43·6 to 45·2 per cent. This last pair of figures, however, obscures an important change in the composition of output within the underdeveloped countries: the share of Asia and Africa increased from 19·7 to 27·0 per cent while the share of Latin America declined from 23·9 to 18·2 per cent. As in the earlier cases of cocoa, coffee and bananas, it appears that Africa has increased her share of the world copper market at the expense of Latin America.

Unfortunately, since the end of the Second World War output in the largest Latin American copper producing country, Chile, failed to grow rapidly and, indeed, production of the higher priced refined copper remained completely stagnant.[7] In the period 1945–49

[4] As can be seen in the regression equation, there is a very close association between the world demand for copper, expressed in thousands of short tons (C) and (M), the index of world industrial output (1958 = 100). The equation covers the 17 years from 1950 to 1966.

$$C = 941{\cdot}09 + 31{\cdot}77\,M$$
$$R^2 = 0{\cdot}98$$
$$t = 31{\cdot}30$$

The elasticity of demand for copper with respect to a change in world industrial output is 0·79.

[5] A Maizels, *op. cit.*

[6] It is generally believed by those well acquainted with the copper industry that at least 20 per cent of world output is produced at a cost greater than US 25 cents per pound, and 45 per cent at a cost greater than 20 cents. The high cost mines are concentrated in the United States, while in Chile no mine produces at a cost in excess of 20 cents.

[7] In 1948 production of refined copper in Chile was 411·8 thousand short tons; in 1966 it was 400·2 thousand tons.

average total output was 4,367 million pounds.[8] It declined sharply to 3,805 million pounds in 1950–54, recovered thereafter and reached a level of 5,501 million pounds in 1960–64. Over the entire period the trend rate of growth of output was less than 1·6 per cent per annum. Chile's share of the world primary copper market declined steadily from 19·6 per cent in 1946 to 12·7 per cent in 1966.

Within Chile the growth of output, in relative terms, came increasingly from the small and medium mines. The large American owned mining enterprises, the so-called 'Gran Minería',[9] produced less copper in 1961 than in 1943. Over the entire period 1943–66 output from the 'Gran Minería' increased only 11·05 per cent, or at a trend rate of less than 0·5 per cent per annum.

Table IV:1

OUTPUT OF THE GRAN MINERÍA, 1943–66
(short tons)

1943	539,475	1955	433,458
1944	540,711	1956	489,338
1945	509,435	1957	480,590
1946	395,359	1958	460,330
1947	450,262	1959	548,494
1948	468,463	1960	528,629
1949	386,686	1961	530,821
1950	380,368	1962	562,855
1951	397,070	1963	559,861
1952	412,274	1964	582,237
1953	358,443	1965	547,384
1954	355,429	1966	596,391

The falling share of the world market, the slow growth of exports, the chronic balance of payments problems, the necessity for frequent and large devaluations and the impact of these upon the already high rate of inflation all gave rise to concern about the performance of the copper sector. Indeed, copper policy was a major issue in the Presidential elections of 1964. The Popular Front advocated nationalization of the American owned Gran Minería, while the victorious Christian Democrats countered this with a proposal for 'Chileaniza-

[8] Production in this chapter is measured in pounds and short tons. (2,000 pounds = 1 short ton = 0·9072 metric tons. One metric ton = 1,000 kilos = 2,205 pounds.)

[9] The 'Gran Minería' is composed of mines which produce 75,000 metric tons or more.

tion' of the copper industry as part of their programme of 'Revolution in Liberty'.

President Frei entrusted the negotiations with the American copper companies to two men—an engineer, Raúl Sáez, and a lawyer with considerable experience in marketing copper, Javier Lagarrigue. After being in office only seven weeks President Frei was able to present to the nation, in a speech on December 21, 1964, a $500 million programme which would expand copper production by 5·6 per cent per annum, raise Chile's share of the world market to 18·9 per cent by 1971, and give Chileans greater control over the marketing and partial ownership of the nation's most important export industry. It is the purpose of this chapter to analyse Frei's copper expansion-Chileanization programme and compare it with the policy advocated by his major opponents.

2 OBJECTIVES

The broad objectives of the copper expansion programme were to increase copper production by 377 thousand short tons,[10] to refine as much of Chile's copper as possible in refineries located in Chile,[11] to reduce imported inputs into the copper sector and increase domestic supplies,[12] and to increase the participation by the State in the production and marketing of copper. To achieve these objectives separate negotiations with each of the foreign companies were conducted.

El Teniente

The heart of the copper expansion programme is the purchase of a 51 per cent controlling interest in the Braden Copper Co., a subsidiary of Kennecott, and the expansion of its mine, El Teniente, the largest underground copper mine in the world and the second largest mine in Chile. The new mixed company is known as the Sociedad Minera El Teniente, S.A.

For purposes of expropriation and compensation Braden was valued at $160 million, although Kennecott originally asked for

[10] This figure refers to the Gran Minería only. It was hoped that increased production from other sources—presumably small and medium mines—would raise total copper output by about 500 thousand tons, but no specific projects were prepared.

[11] It was hoped to raise refining capacity from 302,500 short tons to 770,000 tons.

[12] It is hoped that $12·5 million of import substitution will occur in the copper sector.

$200 million.[13] Chile's share thus came to $81·6 million, which was paid to Kennecott in five annual instalments bearing an interest charge of 4½ per cent.

The value placed upon the Braden Copper Co. has led to considerable controversy in Chile and to this observer it seems clear that much too high a price was paid. At one extreme one could use the book value of the enterprise as the basis for compensation. In the case of Braden in 1963 this would be $65·7 million, 51 per cent of which equals $33·5 million. Critics of this procedure would argue, of course, that the book value of a company bears no relation to its commercial value, particularly since the value of its assets were written down quickly under the legal provisions for accelerated depreciation. On the other hand, it could be argued that the provision for accelerated depreciation, in effect, is an interest free loan from the Chilean government to Braden and allows the company to reduce its tax liability in the present by shifting accounting profits forward into the future. The company should not be allowed to have its cake and eat it too, i.e. use one figure of the value of its assets for tax purposes and a higher figure for purposes of compensation.

At the other extreme, as an upper limit on Braden's value, one could calculate the present value of expected future net profits. This could be described as a strictly 'commercial' criterion.[14] In Braden's case let us assume that net profits remained constant at $12·67 million a year, despite the fact that (a) profits were falling in recent years and (b) it is well known that the mine was running down and in urgent need of additional investment if output was to be sustainined and serious labour problems avoided. The present value of this stream of profits over thirty years, discounted at ten per cent, is $119·82 million dollars, 51 per cent of which equals $61·1 million.

Thus no matter which procedure one uses, Braden was overvalued. Instead of the $180 million agreed upon, the true value of the enterprise was somewhere between $65·7 million and $119·82 million, and the cost to Chile of a controlling interest should have been somewhere between $33·5 and $61·1 million. In other words, Chile paid $20·5—48·1 million too much. If, instead of paying $81·6 million, Chile had paid the average of our upper and lower limits, i.e. $47·3 million, the attractiveness of the 'Chileanization' pro-

[13] See R. Sáez, *Reportaje Efectuado por el Diario 'El Mercurio'*, January 24–30, 1965.

[14] In other words, the government compensates a commercial enterprise for its loss of profits—including those profits due to its oligopoly position.

gramme would have greatly increased and, as we shall see, one's evaluation of the programme would have to be modified.

Be that as it may, it was agreed to expand production by 100 thousand short tons, 68 per cent of which would be refined.[15] Employment during the peak of the construction period was expected to rise by 3,250–4,650 men, but once the expansion programme was finished permanent employment was expected to decline by five men. The investment period was expected to last forty-two to fifty-four months and the cost of the project, including re-housing a large portion of the labour force, was estimated to be $230·241 million.

Rio Blanco

A separate agreement was reached with the Cerro Corp., a group which had not previously been operating in Chile, to establish a mixed enterprise and develop a new mine, Rio Blanco. A new company was formed, Sociedad Minera Andina, with a share capital of $6 million, 25 per cent of which was provided by the Chilean government and the remainder by the Cerro Corp.

The Rio Blanco mine is designed to produce 66 thousand short tons of copper a year and thus, technically, is not part of the Gran Minería at all.[16] Maximum employment during the construction period is estimated to be 2,250 men and the number of new permanent jobs created once the mine is operating at full capacity will be about 850. The gestation period of the project is expected to be 46–48 months and its total cost $89 million.

Anaconda Group

The Anaconda Group is involved in the expansion of three mines.[17] Its major subsidiary, the Chile Exploration Co., owns the largest mine in the country, Chuquicamata, and will invest $99·107 million in its expansion. A second mine, Exótica, in fact an outcrop of the Chuquicamata deposit, will be exploited by a new mixed enterprise, Campañía Minera Exótica. The share capital of Exótica is $15

[15] The engineering details of the project are described in Corporación del Cobre, *Informe 'Proyecto 280', Sociedad Minera El Teniente, S.A.*, October 1966. Henceforth the Corporación del Cobre will be written as CODELCO.

[16] The project is described in CODELCO, *Informe Proyecto Rio Blanco, Sociedad Minera Andina, S.A.*, September 1966.

[17] These projects are described in CODELCO, *Informe: Programa Grupo Anaconda*, December 1966.

million, 25 per cent of which was provided by the Chilean government and the remainder by Anaconda. The total cost of developing Exótica is expected to be $38 million.

Chuquicamata and Exótica combined will increase copper production by 191·5 thousand tons, almost all of which will be refined. In addition, a small mine, El Salvador, will expand its production by 20,000 tons, at an investment cost of $10·3 million. A further $60 million will be spent by the Anaconda Group to construct 5,365 houses for its work force. The entire programme should eventually generate 250–300 new jobs.

Summary

The 'Chileanization' programme will cost over half a billion dollars, a large proportion of which is in the form of foreign exchange. Copper production and sales will rise by about 377 thousand tons and permanent employment should rise by approximately 1,100

Table IV:2

COPPER EXPANSION PROGRAMME

	Investment Requirements (million US dollars)			Increased Output (thousands of short tons)	Increased Employment (persons)
	local currency	foreign currency	total		
El Teniente	120·2	110·0	230·2	100·0	−5
Rio Blanco	41·2	47·8	89·0	66·0	850
Chuquicamata	38·1	61·0	99·1	191·5 (Chuquicamata and Exótica)	250–300 (Chuquicamata, Exótica and El Salvador)
Exótica	15·9	22·1	38·0		
El Salvador	3·8	6·5	10·3	20·0	
Anaconda Housing Project	60·0	0	60·0		
Total	279·3	247·4	526·7	377·5	1,095–1,145

Source: CODELCO.

men. The incremental capital-output ratio of the programme, assuming a sales price of 34 cents per pound of copper, is 2·4:1 and the incremental capital-labour ratio is roughly $47,877.

A summary of the programme is presented in the table above.

3 CAPITAL COSTS

The purpose of this section is to analyse the sources of finance for the expansion programme and to determine how the capital costs are distributed among each of the two most interested parties: the Chilean government and the foreign companies. Where appropriate, future costs have been expressed in terms of present value and a discount rate of ten per cent has been used to do this. All values are expressed in terms of constant (1965) US dollars.

El Teniente

The $230 million project at El Teniente is financed through a series of loans. The most important of these is a loan for $110,116·00 from the US Export-Import Bank (X–M Bank). The loan has a five-year grace period and is repayable over the next fifteen years. The provision for accelerated amortization, however, means that most of the loan will be amortized five years after repayments begin. The rate of interest is 6 per cent on the amount expended and 0·5 per cent on the unused balance. The loan is tied to purchases of US products[18] and is conditional upon the new mixed enterprise offering a management contract to the former Braden managers.

Next, the Chilean government through its agent, CODELCO, will provide two loans with a total face value of $27,482·00. The grace period and repayment period are identical to those of the X–M Bank loan. The rate of interest is 5·75 per cent on the amount used. Finally, Kennecott will provide two loans to El Teniente with a total face value of $92,743·00 on terms similar to the CODELCO loans.

The costs to Chile of the project are reasonably clear. They consist, first, of the value of the equity payments, second, of 51 per cent of the repayment stream of the X-M Bank loan, and third, of the capital provided by CODELCO in the form of loans—all suitably discounted. The present value of this stream of payments is $129·39 million.

The costs to Kennecott are much less clear. Most of Kennecott's contribution to the costs of expansion are more apparent than real. In effect, one asset of the foreign corporation (*viz.* a 51 per cent equity in El Teniente valued at $81·6 million) has been exchanged for

[18] A recent UNCTAD study suggests that tied aid in Chile has raised costs by at least 12·5 per cent above what they need have been. In so far as this general conclusion can be applied to this particular case it implies that the costs of expanding El Teniente are at least 5–6 per cent higher than necessary. (See UNCTAD *Crecimiento, Financiación del Desarrollo y Ayuda: informe sobre créditos condicionados: Chile*, [TD/7/Supp. 8/Add. 1], December 8, 1967.).

another (*viz.* a loan credit to El Teniente of $92·7 million); the former asset yields profit income and the latter yields interest income. Virtually all that has happened is that there has been a change in the composition of Kennecott's assets. No new capital has been raised by Kennecott and no consumption has been forgone by Kennecott's shareholders—except for the $11·1 million difference between the value of the loan and the value of the equity. It is this undiscounted sum plus 49 per cent of the repayment stream of the X–M Bank loan which constitutes the real cost of the expansion programme to Kennecott. The present value of this stream of payments is $39·23 million.[19]

Rio Blanco

The Rio Blanco project is financed through a combination of share capital, debenture issues and foreign loans. The share capital of the new corporation is $6 million, divided between the government and the Cerro Corp. in the ratio of 1:3. Next, there are five series of debentures. Series W has a face value of $10 million, yields 9 per cent per annum and is repayable over five years, beginning eighteen months after production commences. Series X has a face value of $13·3 million, yields 6 per cent and is repayable during the period 1977–87. The Chilean government has subscribed for half of this series and Cerro for the remainder. Series Y has been purchased by CODELCO; it has a face value of $3·3 million, yields 6 per cent and is repayable during 1977–82; these debentures are held by the Cerro Corp. Series U yields 7 per cent a year on a face value of $8 million; it too is held by Cerro and is repayable in two years after production begins. Finally, there are two loans—one from the X-M Bank worth $35,355,000 and another from the Sumitomo Co. worth $10,135,000. The interest rate is 6 per cent and the loans are repayable over fifteen years beginning in 1971, although a clause providing for accelerated amortization means that most of the loans will be paid back after only four years.

The costs to Chile of her participation in the project consist of her share of the equity, fifty per cent of the capital tied up in debentures Series X, the capital for Series Y and 25 per cent of the costs of servicing Series W and the loans from the X-M Bank and the Sumitomo Co. The present value of these payments is approximately

[19] One might wish to dispute the proposition that Kennecott bears a share of the cost of repaying the X–M Bank loan, since all foreign loans used to finance the copper expansion programme are guaranteed by the Chilean government.

$19·35 million. The Cerro Corp. is responsible for all other costs, *viz.* 75 per cent of the share capital, 50 per cent of Series X, all of Series U and Z, and three-fourths of the costs of servicing Series W, the X-M Bank loan and the Sumitomo loan. The present value of these payments is $46·97 million.

Although the distribution of costs in the present project is not nearly as inequitable as in the El Teniente project, it is still noteworthy that the Cerro Corp. is entitled to 75 per cent of the profits although it will incur only 70·8 per cent of the capital costs.

Anaconda Group

Most of the expansion of the Anaconda Group in Chile—*viz.* $133·71 million—will be provided out of retained earnings of the corporation. In addition, the X-M Bank will provide three loans: $21 million for Exótica, $30 million for Chuquicamata and $7·7 million for El Salvador. Lastly, $11·25 million will be provided by Anaconda as share capital to Exótica and CODELCO will provide $3·75 million in equity capital.

The cost to Chile is only her portion of Exótica's share capital and 25 per cent of the interest and amortization charges on the first X-M Bank loan. All other costs of expansion will be borne by Anaconda. The discounted present value of the costs is $6·66 million for Chile and $146·16 million for Anaconda.

4 BENEFITS

The benefits attributable to the copper expansion programme, and their distribution between the Chilean government and the foreign companies, depends upon what assumptions one makes about (i) conditions under the *status quo*, (ii) the time horizon, (iii) the discount rate, (iv) the long-run average sales price, (v) the level of production in each year, (vi) average costs of production and (vii) changes in tax rates and subsidies.

For purposes of analysis we assume the discount rate is 10 per cent, the time horizon is thirty years, and the long-run average sales price is 34 cents per pound. Costs of production are believed to be 17 cents per pound at El Teniente, 20 cents per pound at the Rio Blanco mine and 19 cents per pound on average for the Anaconda group. In the absence of the programme it is assumed that production at El Teniente would have remained constant at 360 million pounds and would have declined after five years to 800 million pounds within the Anaconda group. At the old effective tax rate on profits of 79·3 per cent for El Teniente and 61·3 per cent for Anaconda the receipts of the Chilean government would have been about $122·09 million a year. The net profits of Kennecott and Anaconda combined would have been approximately $59·11 million a year.

The level of output and gross profits after the expansion programme, for each of the three mining operations, are indicated in the table below.

Table IV:3

COPPER EXPANSION PROGRAMME: PRODUCTION AND GROSS PROFITS

	Production (millions of pounds)			*Profits* (millions of dollars)		
Year	El Teniente	Rio Blanco	Anaconda Group	El Teniente	Rio Blanco	Anaconda Group
1	360	0	820	61·20	0	123·00
2	360	0	860	61·20	0	129·00
3	400	0	920	68·00	0	138·00
4	480	0	980	81·60	0	147·00
5	520	134·5	1·165	88·40	18·83	174·75
6	560	155·4	1·225	95·20	21·76	183·75
7	560	150·0	1·225	95·20	21·00	183·75
8	560	145·4	1·225	95·20	20·36	183·75
9	560	123·7	1·225	95·20	17·32	183·75
10	560	125·7	1·225	95·20	17·60	183·75
11	560	117·8	1·225	95·20	16·49	183·75
12	560	113·8	1·225	95·20	15·93	183·75
13	560	124·9	1·225	95·20	17·49	183·75
30	560	124·9	1·225	95·20	17·49	183·75

Source: CODELCO.

In order to increase its participation in the copper industry and induce the foreign companies to expand output the Chilean government, first, acquired partial ownership of three different mines, as already described, and, second, reduced tax rates. As a result of these two changes the government's share in gross profits declined to 72·56 per cent at El Teniente and to 52·6 per cent at Anaconda; the share of profits from the new Rio Blanco mine was set at 55·4 per cent.[20] It was further agreed the rates would remain unchanged for at least twenty years.

In addition, it was agreed with the Cerro Corp. that the Rio Blanco mine would receive a subsidy in the form of cheap electricity from the National Electricity Co., ENDESA. This subsidy is worth $1·4 million a year and is financed by CODELCO, compensating ENDESA, out of its share of Rio Blanco's profits.

The income of the Chilean government after expansion is equal to tax receipts minus subsidies, plus profit income on government's equity holdings plus interest income on loans made to El Teniente and Rio Blanco. This sum minus the government's income under the *status quo* represents the benefits to Chile of the copper expansion programme. These flows, of course, must be discounted in order to obtain the present value of the benefits.

Applying this procedure, it turns out that the benefits to Chile of the Chileanization scheme are as follows:

El Teniente	$134.48 million
Rio Blanco	59.00 million
Anaconda Group	118.69 million

The benefits to the foreign companies consist of the discounted profit and interest income generated by the programme minus discounted profits under the *status quo*. These benefits are as follows:

Kennecott	$129.00 million
Cerro	57.31 million
Anaconda	295.89 million

Although the present value of the benefits to Chile of the expansion programme over the entire thirty years is positive, in the early years

[20] The profits and taxes received by the Chilean government are calculated as follows:

El Teniente:	profits tax	20·00%
	51% share of remaining 80%	40·80
	Additional tax of 30% on income of foreign shareholders	11·76
		72·56
Rio Blanco:	profits tax	15·000%
	25% of remaining 85%	21·250
	Additional tax of 30% on foreign shareholders	19·125
		55·375
Anaconda Group:	Exótica (same as Rio Blanco)	55·375%
	El Salvador profits tax	50·000
	Chuquicamata profits tax	52·500
	Unweighted average	52·625

of the scheme the benefits are negative. That is, if one takes a very short-run view the *status quo ante* is preferable to the Chileanization scheme. As can be seen in the table below, as far as Chile is concerned, El Teniente has negative benefits in the first two years, Rio Blanco yields no return until the fifth year and the Anaconda Group provides negative benefits for the first four years. Chile's total benefits are negative in the first three years. In the fourth year total benefits are positive, but they are insufficient to compensate for the losses of the first three years. It is only from the fifth year onwards that cumulative total benefits become positive.

Table IV:4

PRESENT VALUE OF CHILE'S BENEFITS: FIRST FIVE YEARS
(millions of dollars)

Year	El Teniente	Rio Blanco	Anaconda Group	Total	Cumulative total
1	−3·66	0	−9·70	−13·36	−13·36
2	−2·90	0	−6·58	−9·48	−22·84
3	1·51	0	−3·12	−1·61	−22·45
4	8·31	0	−2·12	6·19	−18·26
5	10·68	6·01	9·77	26·46	8·20

At no stage, it should be added, are the benefits to the foreign companies negative. With the exception of the new Rio Blanco mine, the foreign companies—beginning in the very first year—make larger net profits under the Chileanization programme than under the *status quo ante*.

5 A GENERAL ASSESSMENT

We are now in a position to make a broad evaluation of the copper expansion programme from the point of view both of the Chilean government and the foreign companies. This is, of course, a theoretical exercise—the outcome of which is sensitive to the assumptions we have made. The level of benefits, for example, depends upon what the long-run price of copper is assumed to be. We assume it will be 34 cents per pound; the government, in making its original calculations, assumed it would be 29 cents per pound; in practice it may be above or below this range.[21] In all cases we have tried to be conservative yet realistic, i.e. the assumptions were chosen so as not to overstate the benefits.

The distribution of benefits between the government and the companies is not highly sensitive to small changes in the assumptions. A large increase in gross profits, however (for instance, due to a substantial increase in the world price), would lead to a redistribution of benefits in favour of the government. This is because at present Chile receives slightly less than 40 per cent of the total benefits of the programme, whereas her share of any increase in benefits is considerably greater than 50 per cent. Since Chile's marginal benefits are greater than her average benefits, the country will gain more than proportionately from any rise in gross profits.[22] Conversely, Chile will lose more than proportionately from any fall in profits. In other words, Chile will gain from any events which sharply increase the demand for copper, e.g. wars, or reduce supply elsewhere, e.g. political disturbances in Africa.

In the table below the benefits and costs are summarized for each project. The overall attractiveness of each project is indicated in two ways: first, by dividing benefits by costs and obtaining a benefit-cost ratio and, second, by subtracting costs from benefits and obtaining the net present value. If the benefit-cost ratio is greater than one or, alternatively, the net present value is greater than zero, then the contribution of the project to the net income of the government or the companies, as the case may be, is at least positive. Since we are using a discount rate which we believe reflects the marginal productivity of capital in Chile, although in a very rough and ready way, any project with a benefit-cost ratio greater than unity should be undertaken.

The essence of the Chileanization programme is the partial nationalization of El Teniente; the other elements of the programme comprise little more than a reduction in tax rates and the purchase of a 25 per cent equity in two small new mines. It is on El Teniente, therefore, that one must concentrate in trying to assess the government's policy.

[21] The current producers' price is 49 cents per pound and the long-run price is believed to be about 38 cents per pound. The latter is roughly consistent with our assumption of 34 cents per pound, since we are dealing with constant 1965 dollar prices.

[22] Assuming costs of production remain unchanged a 1 cent per pound rise in copper prices and company profit margins will increase gross profit from the completed expansion programme by $7·55 million. Chile will receive roughly 58·4 per cent of this, or $4·4 million, and the companies will receive $3·15 million.

Table IV:5

SUMMARY OF INVESTMENT ANALYSIS

Project	Chilean Government	Foreign Companies
El Teniente		
Benefits (B)	134·48	129·00
Costs (C)	129·39	39·23
B/C	1·04	3·29
B-C	5·09	89·77
Rio Blanco		
B	59·00	57·31
C	19·35	46·97
B/C	3·05	1·22
B-C	39·65	10·34
Anaconda Group		
B	118·69	295·89
C	6·66	146·16
B/C	17·82	2·02
B-C	112·03	149·73

It is noteworthy that although the benefits of expanding production at El Teniente are divided almost equally between the Chilean government and the Kennecott Corp., the costs to Chile of obtaining these benefits are over three times greater than the costs to Kennecott. In fact, Chile's benefit-cost ratio is almost exactly one. In other words, the *net* benefits to Chile of participating in the El Teniente project are very close to zero. Chile's negotiators wanted their country to participate more fully in the copper industry and this they were able to achieve. But Chile's participation involves absorbing three-fourths of the costs while allowing Kennecott to enjoy half the benefits.

The terms under which El Teniente was partially nationalized were so generous that Kennecott ended the process of negotiation with a higher benefit-cost ratio than either of the other two foreign companies and Chile was left with practically no net benefits at all. Unless the Chilean people enjoy pyschic income from the mere fact of legal majority ownership of the mine, the agreement with Kennecott appears to be a disaster from Chile's point of view.

One reason for this is the high cost of compensation. It has already been argued that the government paid far too much to acquire a controlling interest. If, instead of paying $81·6 million for 51 per cent of Kennecott's subsidiary, the government had paid in cash our mean estimate of what the enterprise was worth, i.e. $47·3 million, the benefit-cost ratio would have risen to 1·26 and the net present value would have increased to $27·95 million. This still would not have been a very attractive proposition, but it would have been much better than what actually occurred.

The Rio Blanco project, in contrast, is much more satisfactory. Chile's benefit-cost ratio is quite high and the rate of return on Cerro's investment is kept down to a level which, perhaps, is not too unreasonable. Even so, Chile is bearing a slightly disproportionate share of the capital costs, since the government has only 25 per cent of the equity yet is contributing 29·5 per cent to the costs of developing the mine.

Chile seems to have benefited considerably from the agreement with Anaconda: her capital costs are negligible and the increased government revenues are substantial. As a result the benefit-cost ratio is unusually high, *viz.* 17·82. In this case, however, the ratio is misleading, since most of the benefits consist of additional tax receipts and have almost nothing to do with CODELCO's investment in 25 per cent of Exótica's equity. In other words, if the government had persuaded Anaconda to expand production simply by reducing tax rates—and this certainly was possible—Chile would have incurred no capital cost and the benefit-cost ratio would have been infinitely high. Yet this would not in itself constitute a justification for such a policy.

One approach to the problem would be to say that Anaconda's benefit-cost ratio should be no higher than Cerro's benefit-cost ratio on the Rio Blanco project, since the opportunity cost of capital to both foreign companies should be very similar. Applying this criterion it follows that Anaconda's benefits should be reduced, i.e. Chile's tax receipts should increase, by $117·55 million in order to lower the benefit-cost ratio from 2·02 to 1·22. Such a policy would double Chile's benefits from the Anaconda projects while leaving the parent company no worse off than one of its major competitors.

In summary, as a policy designed to maximize Chile's income from her copper resources the Chileanization programme must be considered a failure. In both the El Teniente and Rio Blanco projects, particularly in the former, Chile's costs were excessive, and in the Anaconda project her benefits were deficient. It seems clear from this experience that the Chilean negotiators were at a severe disadvantage when bargaining with the foreign companies. They were subject to strong pressure to treat the companies 'fairly'; they were handicapped by their ignorance of many details of the industry and their lack of competent economic advice; and for internal political reasons they were urged to reach agreement quickly. The partnership that was formed with the companies has been decidedly unequal and as this becomes apparent to the majority of the population it is likely to become an embittered partnership as well.

6 THE NATIONALIZATION ALTERNATIVE

In the election campaign of 1964 the Popular Front stated that it was its policy to nationalize the foreign owned copper companies. In this section we propose to compare this policy of nationalization with that of 'Chileanization'.

Unfortunately, the Popular Front, like the Christian Democrats, did not indicate in detail how it would implement its policy. Thus in assessing nationalization as a policy we must make certain assumption about what would have happened if Senator Allende had been elected President. Our basic assumption is that the foreign companies would have been nationalized in late 1964 and compensated on the basis of the book value of their 1963 capital assets. The cost of this would have been $372 million, distributed among the three subsidiaries as follows:

Andes Copper Mining Co. (El Salvador)	$93.4 million
Braden Copper Co. (El Teniente)	$65.7 million
Chile Exploration Co. (Chuquicamata)	$213.1 million
	$372.1 million

We further assume that this would have been financed with bonds given to the companies. Specifically, the companies would make a twenty year forced loan to the government, payable in annual quotas which include straight-line amortization plus an additional 6 per cent of the yearly amortization payment; there would be no grace period. The annual cost of nationalization, i.e. amortization of the forced loan plus interest, thus would be $19·73 million. The present value of this stream of yearly payments, discounted at 10 per cent, is $168·02 million. This is the cost of acquiring the companies.

Next, we assume that the efficiency of the industry is maintained and that all the profits of the copper industry, including those due to the expansion programme if this is implemented, accrue to the Chilean government. Some observers might object that it is most unlikely that output and profit margins would remain unaffected by nationalization. This is too pessimistic a view, however. There already are a number of industries in Chile wholly or partially owned by the state, and several of these are quite efficient, e.g. the national airline (LAN), the national petroleum company (ENAP), the national electricity company (ENDESA), and the steel industry (CAP)—a mixed enterprise. Furthermore, the copper industry is almost entirely staffed and managed by Chileans. At El Teniente in 1964, for example, there were only fourteen foreigners on the payroll, i.e. 0·02 per cent of the staff.

The additional income that would be received by the government, assuming efficiency is maintained, is equal to gross profits minus taxes under the old system, or the net profits of the foreign companies. This additional annual income is indicated in the next table.

Table IV:6

PROJECTED BENEFITS OF EXPROPRIATION
($ million)

Year	Projected Benefits
1	60.27
2	60.56
3	61.14
4	63.46
5	60.79
6	59.11
:	:
30	59.11

The present value of the above stream of benefits, discounted over thirty years at 10 per cent, is $565·03 million. The benefit-cost ratio of a nationalization programme which does not include expansion of the copper industry is, thus, 3·36 and the net present value is $397·01 million. In other words, the benefit-cost ratio of nationalization is higher than that of the Chileanization programme (3·36 *v.* 2·01) and the net present value of nationalization is over 150 per cent greater than the net present value of the current policy ($397·01 million *v.* $156·77 million).

The analysis can be extended further, however. There is no reason to suppose that the Popular Front would have been content to nationalize the copper industry and then not expand it. For purposes of comparison let us assume the same physical output targets would have been adopted and the costs of expansion—*viz.* \$526·65 million —remained unchanged. Let us also assume that foreign loans would not have been available and that the entire cost of investment, spread over seven years, would have been financed from current domestic savings.[23] The phasing of this investment is shown in the following table.

Table IV:7

THE INVESTMENT SEQUENCE
(\$ millions)

Year	Investment Cost
1	17.9
2	95.0
3	132.0
4	121.0
5	103.0
6	43.0
7	14.7

The present value of the stream of investment costs, discounted at 10 per cent, is \$372·3 million. The benefits of the expansion phase of the nationalization programme are the sum of the benefits received by the government and the foreign companies in Table 5 minus income received in the form of interest payments on loans from CODELCO, Kennecott and Cerro. These benefits are equivalent to \$748·36 million. The benefits and costs of the entire nationalization and expansion programme are listed in the following table.

It is obvious from these results that a programme of nationalization and domestically financed expansion of production is superior to any other policy considered. Indeed, the net present value of this scheme is 4·9 times greater than the net present value of the Chileanization programme. As a theoretical exercise there is no doubt that the espoused policy of the Popular Front, as it has been interpreted here, is more to the country's advantage than the Chileanization programme. Whether such a policy could have been implemented even if the Popular Front had won the election is quite another matter.

[23] Since no part of the investment is being financed by tied loans from the X-M Bank it would be reasonable to assume that capital costs would be somewhat lower. As suggested earlier, it might have been possible to save 5–6 per cent of the total cost by relying on untied sources. To simplify the analysis, however, we assume that total costs are the same under both policies. Hence to some extent we are understating the net benefits of nationalization.

Table IV:8

BENEFITS AND COSTS OF A PROGRAMME OF NATIONALIZATION AND EXPANSION
($ millions)

	Benefits	Costs
Nationalization	565.03	168.02
Expansion		
El Teniente	227·61	372.30
Rio Blanco	106.17	
Anaconda Group	414.58	
Total	1,313.39	540.32

B/C = 2.43
B-C = $773.07 million

7 IMPLEMENTATION OF THE PROGRAMME

In presenting its programme in December 1964 the government assumed the engineering works would commence by July 1, 1965. As might have been expected, however, the Chileanization scheme provoked considerable controversy in the Congress and the decrees authorizing the investments were not issued until late 1966 or early 1967.[24] Thus the initiation of the programme was delayed 17–20 months.

Since the signing of the decrees, however, the work has progressed rapidly. The investments at El Teniente are ahead of schedule and if the present pace continues the work should be completed in late 1970, nearly one year ahead of the plan. Exótica, too, is advancing rapidly and may be in operation twelve months earlier than anticipitated. The other projects are developing more or less as planned, although the expansion at Chuqicamata appears to have been delayed slightly. In general, progress has been very satisfactory. It is always possible, of course, that bad weather or strikes will disrupt the investment process, and one small dark cloud already is on the horizon: estimates of the capital costs at Exótica have been revised

[24] The dates of the relevant decrees are as follows:
El Teniente, March 1, 1967;
Rio Blanco, December 9, 1966;
Chuquicamata, December 23, 1966;
Exótica, February 13, 1967;
El Salvador, December 23, 1966.

upwards from $38 million to $45 million, an increase of over 18 per cent.

As of December 31, 1967, actual investment, expressed as a percentage of total authorized investment, was as follows:

El Teniente	5·3 per cent
Rio Blanco	39·2
Exótica	39·6
El Salvador	53·7
Chuquicamata	26·8

Of the total programme, including the Anaconda housing scheme, 15·8 per cent was completed by the end of 1967.[25]

Output and exports will increase slightly in 1969 and rise rapidly thereafter. The value of exports from the Gran Minería, however, will rise only fractionally because of the expected decline in world prices.

Table IV:9

PROJECTED EXPORTS OF THE GRAN MINERÍA

	Quantity (million lbs)	Price (cents/lbs)	Value ($ millions)
1968	1,199	46·1	553·0
1969	1,286	36·9	474·4
1970	1,547	33·1	512·1
1971	1,795	33·1	594·2
1972	1,815	31·1	564·4

Source: ODEPLAN. The price estimates are for 'net' prices, i.e. exclusive of freight and insurance charges.

The prices projected in the above table are highly conjectural, of course. One can be reasonably certain that prices will fall in the near future, but by how much and how fast is unknown. Many of the forces responsible for the extraordinarily high prices of the past few years have disappeared or have become weaker. Stability in the Congo has increased, Zambia has adjusted—although only partially—to the shock caused by Rhodesia's unilateral declaration of independence, and the 8½ months' copper strike in the United States has been resolved. Thus output in the short-run should recover from

[25] CODELCO publishes periodic progress reports on the expansion programme, although the data are not presented in the most useful form. For the latest information, see CODELCO, *Desarrollo de los Programas de Expansión de la Minería del Cobre*, Situación al 31.12.67, Informe No. 4, April 1968.

these temporary disturbances. Moreover, in the next few years additional output will be placed on the market as a result of the expansion programmes occurring not only in Chile, but in Canada, Peru and Australia as well. In other words, supply increases should lead to lower copper prices. In addition, slower growth in Europe, a pause in the violence of the Middle East and possible de-escalation of the war in Vietnam as a consequence of the Paris negotiations may lead to a reduction in world demand. This tendency for demand to fall, however, may be offset for a while by the efforts of US corporations affected by the strike to rebuild their stocks.

Hence the short-run outlook is for falling copper prices. The long-run outlook is even more uncertain. It is possible that our assumptions about costs and prices may prove to be incorrect, i.e. the profit margin in Chile will be squeezed. Indeed, the less optimistic assumption used by CODELCO in its current calculation—*viz.* that the long-run price of copper in terms of 1968 constant prices will be 35 cents a pound—may turn out to be closer to the truth.

8 CONCLUSION

The copper expansion programme obviously is better than nothing. Chile's net gains from the whole programme are positive and, hence, the new policy is better than the *status quo*. Since we used a discount rate of 10 per cent it is rather unlikely that the returns on investments in other sectors of the economy would have been more attractive than investment in copper. On the other hand, when the various components of the Chileanization programme are examined it becomes clear that many opportunities were lost in the negotiating process.

The keystone of the present policy is the partial nationalization of El Teniente. Yet the terms on which this occurred are such that Chile bears most of the costs while Kennecott receives a disproportionate share of the benefits. As a consequence the net gains to Chile, assuming the price of copper is 34 cents a pound, are virtually zero. The Anaconda group has done almost as well as Kennecott; they have made almost no concessions to the desire of Chileans to regain at least partial ownership of their copper industry, they have benefited from a reduction in tax rates, and they are now about to enjoy an unusually high return on their investment. Since three-fourths of all of Anaconda's assets are located in Chile, it is surprising that the government was unable to strike a tougher bargain. Only in the negotiations with the Cerro Corp. was something approaching parity achieved, yet even here Chile's share of the profits are slightly less than her share of the costs.

At every point in the negotiations—from the value placed upon the Braden Copper Co., to the level of taxation on the profits of the nation's richest resource—the foreign companies were favoured. Even today, after four years of a government committed to a 'Revolution in Liberty', the foreign owned copper companies are given privileges not enjoyed by Chilean businessmen, e.g. the right to import equipment duty free. The nascent entrepreneurial groups in Chile are not given protection from foreign businessmen; they are not even allowed to compete on equal terms; instead they are placed at a disadvantage and subjected to unequal competition. Naturally, the Sociedad de Fomento Fabril complains.

The proposed Compañía Explotadora Cordillera is the latest example of a government initiative which bends over backwards to favour foreign interests. For many years the Anaconda Co. has claimed the mineral rights to huge tracts of land in Chile. Most of this terrain has never been explored and none of it has been exploited. Anaconda has simply kept this land as a potential reserve, maintaining her monopolistic position and preventing others from developing the possible mineral resources located there—much as the latifundistas have monopolized the top-soil rights.

Instead of nullifying these mineral rights so that the land would revert to the State the government has decided to establish a new company—49 per cent of the capital to come from CODELCO and 51 per cent from Anaconda—to search for mineral deposits on land belonging, first, to Anaconda and, much later, the Chilean government. If minerals are found on land belonging to Anaconda, and studies show that these deposits can be exploited with profit, a new mining company would be formed, the ownership of which would be divided between Anaconda and the government in the ratio of 2:1. If the ore deposits are found on government land the Chilean government would hold two-thirds of the equity and Anaconda one-third. Thus Anaconda, instead of being penalized for monopolizing mineral rights and not developing them, is invited by the government to become the senior partner in an exploration company and in the new mining companies that may subsequently be established. Chile, instead of recovering her land and exploiting her own natural resources, becomes a junior partner to a foreign company.

It was shown, in section 6, that nationalization would have been a superior policy to the Chileanization programme. The present policy cannot immediately be reversed, however, because Chile has become ensnared in a series of international loan, tax and management agreements to which the incumbent administration is committed. The best that can be done now is to prepare the way for substantial changes when a new government comes to power in 1970.

It would be in Chile's interests to grant accelerated amortization rights and thereby reduce the book value of the copper companies' assets. In the short-run, of course, this would reduce accounting profits and hence the government's income, but this could be offset in part through the unanticipated high price of copper and through higher tax rates. (Since the agreements preclude an increase in tax rates for twenty years, the later would have to be disguised. This could be done quite easily, e.g. by exchange rate manipulation.) Once a new government is in power and a large proportion of the expansion costs are amortized, the Gran Minería should be purchased by the State at its book value. Unless this or similar policies are followed Chile will find that most of the additional benefits generated by the expansion programme will leak abroad in the form of profit repatriation.

CHAPTER V

Inflation and Exchange Rate Policy

With the exception of Ecuador and Venezuela, all nine of our Spanish American countries have experienced persistent balance of payments pressure and a consequent need for periodic currency devaluations. An important factor until fairly recently has been the tendency for several primary commodity prices to decline and, as the prices of imports of manufactured goods rose, a tendency for the terms of trade to deteriorate even faster. The low rate of growth in the United States up to 1964 caused the demand for Spanish American exports to grow slowly. On the other hand, the supply of the region's exportables (which, relative to alternative activities, are very profitable to produce) has, in some cases, increased substantially. These two factors combined tended to reduce export prices until about 1962.

Latin America as a whole tends to have a surplus on trade account. That is, exports of goods from the region exceed imports valued at f.o.b. prices: in 1962, for instance, exports were 16·5 per cent greater than imports. A major problem is that payments for services of various sorts absorb over 60 per cent of the foreign exchange earnings. Half of the payments for services are accounted for by profits repatriated abroad and servicing of the foreign debt.[1] Clearly it is in the interests of the region to try to reduce its deficit in services. Profit repatriation and the high average level of payments for invisibles, however, can account for only a relatively small portion of the pressure on the balance of payments, although a reduction in profit leakages and invisible exports would go a long way toward correcting the balance of payments position.

Nonetheless, by far the most important explanation of the balance of payments difficulties is to be found in the acute domestic inflation in Spanish America. The table below presents some data on the cost of living in the nine countries since 1953. The base year for the time series is 1958.

[1] A. G. Frank, '*Servicios Extranjeros o Desarrollo Nacional?*', *Comercio Exterior*, (Mexico), February 1966.

Table V:1

INFLATION IN SPANISH AMERICA, 1953–66
(1958 = 100)

	1953	*1955*	*1958*	*1961*	*1963*	*1966*	*1967*
Argentina	36	45	100	309	491	1020	1310
Bolivia	5	17	100	144	151	184	194*
Chile	15	45	100	167	274	632	747
Colombia	69	71	100	121	164	239	258
Ecuador	96	100	100	106	115	130	135
Paraguay	48	71	100	141	146	158	160
Peru	74	81	100	131	148	209	230
Uruguay	56	66	100	237	317	1230	2540†
Venezuela	96	96	100	106	106	109	108

Source: International Monetary Fund.

* Second quarter 1967.
† Third quarter 1967.

Any policy designed to relieve the pressure on the balance of payments must take the persistent price increases into consideration. If one accepts inflation as a datum a once-for-all solution, such as devaluation, is impossible. If, on the other hand, one views inflation and a chronic payments deficit as two aspects of the same problem, it may become possible to design a policy mix to combat both difficulties simultaneously—by attacking their common cause. In accordance with this view, in Part 1 of this chapter we develop a macro-economic theory of inflation applicable to Spanish America. In Part 2 we go on to present a more disaggregated model which allows us to consider in greater detail the social and sectoral forces at work. Finally, in Part 3 we examine the implications of our theory of inflation for exchange rate policy.

1 GROWTH DISEQUILIBRIA

It was argued in Chapter I that the economies of Spanish America are disjoined and poorly articulated. At the macro-economic level this disjointedness is reflected in a structural disequilibrium such that the process of income generation leads to the simultaneous presence of unemployment and inflation. In other words, strong inflationary tensions are inherent in the system; they are created by growth disequilibria between, on the one hand, the productivity of labour and the savings ratio and, on the other hand, between the rate of growth of the work force and the capital-labour ratio. Precisely how this occurs will be explained below.

A simple Growth Model

In a closed economy the rate of growth of national income (q) may be expressed as the product of the average propensity to save (s) and the productivity, or effectiveness, of investment (e). Symbolically, $q = se$. Given the savings ratio, it is obvious that the greater the effectiveness of investment—or, to say the same thing, the lower the incremental capital-output ratio—the faster will be the rate of growth.

Investment, as is well known, has a dual nature. One the one hand it is a component of total demand. On the other hand it contributes to increasing the productive capacity of the economy.

Keynes concentrated on the demand side of investment. He divided aggregate demand between consumption and investment expenditures. Investment was treated as an independent variable and consumption was assumed to depend largely upon the level of income. Thus a precise relationship between investment and consumption was established. Knowing the consumption function and the increment to investment one could predict the increase in demand. This is the well-known multiplier analysis: the rate of growth of demand is a positive function of the rate of growth of investment; the rise in demand is equal to the increase in investment multiplied by the reciprocal of the marginal propensity to save.

The effect of investment on the supply of goods depends upon its effectiveness. Since 'e' represents the increase in capacity output created by investment (I), the rise in the output of goods equals Ie.

Domar recognized that the economy would be in equilibrium, in a dynamic sense, only when the increase in demand generated by a given change in investment was exactly offset by an equivalent increase in the supply of goods and services.[2] Assuming the marginal and the average propensity to save are the same, this will occur when $\Delta I/s = Ie$. There would then be some rate of increase of investment which, if maintained, would result in a sustained growth path. The requisite proportional rate of growth of investment is the product of the marginal propensity to save and the effectiveness of investment, i.e. $\Delta I/I = se$. Thus, once again, it can be seen that the effectiveness of investment, or its inverse, the incremental capital-output ratio, plays a major role in determining the rate of growth of income.

[2] E. Domar, 'Expansion and Employment', *American Economic Review*, March 1947.

Let us examine briefly this important ratio. It is sometimes imagined that the capital-output ratio is an independent entity bearing no relationship to other economic phenomena. This, however, is not the case. In fact, it can readily be seen that the output-capital ratio is the product of the productivity of labour (p) and the labour-capital ratio. If we use (i) to designate the capital intensity with which goods are produced, i.e. the capital-labour ratio, then the output-capital ratio may be represented as $e = p/i$. As the capital-labour ratio rises, the output-capital ratio falls; but as the productivity of labour rises, the output-capital ratio will increase.

Our growth equation can now be written in a somewhat more useful form: $q = s \, . \, p/i$.

If we make an additional, and perhaps rather crude assumption that the proportional rate of growth of the potential labour force (l) is the same as the rate of growth of the population, we can write an expression for the rate of growth of *per capita* income (y) which includes the four structural elements of a Spanish American economy.

$$y = s \cdot p/i - l$$

These four elements—the productivity of labour, the savings ratio, the rate of growth of the labour force and the degree of capital intensity of production—will be analysed individually to determine their impact on development. These structural elements will then be combined in a simple model to demonstrate the way in which they interact to produce inflationary growth and unemployed labour.

Productivity of Labour

Spanish America, like all underdeveloped countries, is characterized by an extremely low level of labour productivity. There are several explanations for the low output of labour in these countries—inadequate supplies of complementary factors of production, e.g., a lack of capital, technical ability and knowledge; poor climate and infertile soils; ill health, deficient nutrition, a lack of physical energy and mental vigour; general poverty, illiteracy and high rates of absenteeism. Many of these elements interact with one another. For example, productivity is low because nutrition is poor and nutrition is inadequate because productivity is low.

In order to produce a rise in the standard of living it is essential to increase the productivity of labour, but because capital is scarce the direct approach of introducing capital intensive methods of production on a very wide scale usually is not possible. Other methods will have to be found.

One possible, although only partial solution consists of a redistribution of certain types of consumption goods towards the poorer members of society. Such an approach is attractive not only on grounds of social justice but also on its strictly economic merits. For a large portion of the population in Spanish America diets are deficient; this is particularly true in the altiplano of Peru and Bolivia, although conditions are nearly as bad in the other countries, especially Ecuador, Colombia and Venezuela. Moreover, standards of housing are low, clothing is poor, morale is low, disease is rampant and public medical services are inadequate. Excluding Argentina, the number of inhabitants per doctor in Spanish America varies between 1,130 in Uruguay and 4,500 in Peru. (Most doctors, of course, are concentrated in the capital and in the larger cities.)

Table V:2

SOCIAL INDICATORS

	Per Capita calory intake	Literacy rate (per cent)	Infant mortality (per 1000 live births)	Hospital beds (per 100,000 population)
Argentina	3,220	91	60	610
Bolivia	2,010	32	99	190
Chile	2,610	84	114	420
Colombia	2,280	62	83	270
Ecuador	2,100	68	90	230
Paraguay	2,400	68	84	220
Peru	2,060	61	84	250
Uruguay	3,030	91	42	640
Venezuela	2,230	80	46	350

Source: Agency for International Development.

Increasing *per capita* consumption expenditure on low-cost housing, better diets, essential clothing and public health will raise levels of energy, health and morale. For example, 'in Costa Rica, during the construction of the Pan American Road, work performance of each person—measured by the loads that were individually moved daily—was increased fourfold by the addition of meat, fruit and vegetables to the scanty rations of rice and beans, and through the organization of collective dining-rooms that improved supply and service'.[3]

[3] J. F. Patino, 'Investment in Health', in S. H. Robock and L. M. Soloman, eds., *International Development 1965*, New York, 1966, p. 43.

In other words, certain marginal consumption expenditures may have positive growth inducing effects. The productivity of labour may increase from two sources: first, time lost from work due to illness or dissatisfaction may decrease and, second, a given labour force may be able to work somewhat harder. Strikes and slowdowns may be reduced and, furthermore, improved morale may mean that labour will be less reluctant to have management rationalize production methods. It is possible that greater co-operation from labour would lead to important institutional changes which would increase output. These might include the wider use of shift working, a lengthening of the customary working day in urban areas, a reduction of the lunch period and afternoon siesta, the elimination of some holidays or the postponement of the age of retirement.

Consumption of educational services—increased literacy; general education, but particularly technical and agricultural training; books, magazines and newspapers; research programmes, and libraries—will have a three-pronged impact. First, labour's productivity and its capacity for change may be improved directly. Second, the stock of knowledge will increase. New technology immediately relevant to the conditions of Spanish America will appear. This will increase the effectiveness of *net* investment. Third, the technological innovations resulting from increased consumption of educational services will also have a large impact on *replacement* investment. Replacement of depreciated capital generally does not involve merely a duplication of the old equipment; normally it takes the form of introducing similar *but improved* machinery which incorporates the technical developments which have occurred since the time of the original investment. Thus, measures which increase the stock of knowledge can have an important effect on growth.

It sometimes appears that we have become the victims of our Keynesian categories and place too much emphasis on the division of output between consumption and investment goods. Implicit in this division is the view that only investment contributes to growth. It is very doubtful that this is correct. Some investment, e.g. in urban real estate, may contribute relatively little to growth, while some consumption may have a considerable impact on growth. Thus the composition of consumption between those goods and services which contribute to faster growth and those which do not should also be considered, although the dividing line between the two classes of goods will always be shifting and rather vague. The essential point, however, is that quite a bit can be done to raise productivity which uses little capital per man.

A more equal distribution of consumption also has its political dimension. That is, one can justify a redistribution of consumption goods, especially food, to the poorer classes not only for humanitarian reasons or because it will raise productivity but also on the grounds that a more egalitarian society would increase the internal political and economic cohesion of the Spanish American countries. The class distinctions in Spanish America are not, by and large, based on race; they are based essentially on economic divisions arising out of the extremely unequal distribution of income and wealth.

This unequal income distribution in an environment of slowly rising or stagnant *per capita* income has had the effect of creating and accentuating nutritional problems. Since the late 1940s *per capita* consumption of proteins and calories apparently has declined in Colombia and Argentina. *Per capita* consumption of meat has declined in Uruguay, of cereals in Peru and Argentina, of cereals and meat in Chile, and of cereals and sugar in Venezuela.[4] These nutritional problems, in turn, are reflected in the low average productivity per worker, the sharp resentment of the poorer classes and the low regard of the poor for the political institutions of their country.

The Savings Ratio

The savings ratio, together with the output-capital ratio, is a major determinant of the rate of growth of output. Yet one of the striking features of Spanish America, with the exception of Venezuela, is that the proportion of income saved is very low. As can be seen in the table below, gross domestic savings varies from 9 per cent of GDP in Bolivia to about 23 per cent in Venezuela; net domestic savings range from roughly 3 per cent to 13 per cent of GDP. Even these ratios, low as they are, probably give a false impression of the extent to which consumption is being restrained in order to accelerate development, as part of the savings leaks abroad as capital flight and an unduly large proportion of the savings kept at home is invested in real estate and luxury construction and is more akin to luxury consumption of durable goods than to genuine capital formation.

Various techniques have been advocated to raise the low propensity to save and accelerate the growth of capital. The creation of savings and credit institutions to encourage small savers is recommended everywhere (although this may only alter the composition rather than the quantity of savings); squeezing agriculture through confiscatory taxation policies has been important historically; more

[4] Although consumption standards have fallen in Argentina it must not be forgotten that they were quite high to begin with.

recently external grants and loans have assumed some importance as a source of finance for investment; many countries have used inflationary financing, although this method is seldom endorsed by economists; and some have relied on redistributing income towards the business sector in order to increase profits.

Table V:3

THE SAVINGS RATIO

	Gross domestic savings	Net domestic savings
	(per cent of GDP, 1962–64)	
Argentina	18[a]	n.a.
Bolivia	9	3
Chile	11	3
Colombia	16	6
Ecuador	14	9
Paraguay	15[b]	n.a.
Peru	23[b]	n.a.
Uruguay	13	n.a.
Venezuela	23	13

Sources: UN, *World Economic Survey 1965*, 1966; ECLA, *Economic Survey of Latin America 1964*, 1966.
a = Average 1950–59
b = Gross investment as per cent GDP, 1962–64.

Table V:4

THE DISTRIBUTION OF INCOME

	High Incomes		Low incomes	
	per cent of families	per cent of personal income	per cent of families	per cent of personal income
Argentina	n.a.	n.a.	50·0	20·5
Chile	12·5	48·1	54·9	15·7
Colombia	5·0	41·6	60·0	31·4
Ecuador	1·2	17·0	78·1	54·7
Peru	n.a.	n.a.	50·0	12·5
Venezuela	12·0	49·0	45·0	9·0

Source: ECLA, *Economic Bulletin for Latin America*, Vol. VII, No. 2, p. 230; ECLA, *Estudios Sobre La Distribucion del Ingreso en America Latina*, E/CN 12/770/Add 1., April 21, 1967; S. Kuznets, *Economic Development and Cultural Change*, January 1963.

The experience of Spanish America suggests quite clearly that inflation and a very unequal distribution of income have had no effect on raising the savings ratio to the level necessary to sustain rapid growth. The scanty information available seems to indicate that the upper ten per cent in the income distribution receive about 45 per cent of the personal income, yet this very wealthy minority seldom saves enough to raise the net savings rate above 10 per cent of the national income.

The wealthy members of the community appear to have an unusually high propensity to consume, and a large proportion of their expenditure is on imported goods. Sternberg has studied the consumption and savings behaviour of twenty large landowners in the Central Valley of Chile.[5] The average size of agricultural properties of the latifundistas studied by Sternberg was 11,540 ha., yet they consumed 83·6 per cent of their disposable income; the median level of consumption was 99 per cent. Were it not for one landowner who saved 89 per cent of his disposable income, however, the average consumption would have been even higher: eliminating this exceptional case, the mean savings of the remaining nineteen was 6·8 per cent of their disposable income.

In Colombia, despite the fact that 5 per cent of the families account for about 42 per cent of the personal income, domestic savings recently have remained very low: in 1962 they were 8·9 per cent of the national income and in 1963 they fell sharply to a low of 5·0 per cent.[6] Inadequate private savings are not compensated by high public savings out of taxation. Tax receipts of the public sector are lower in Colombia than in any other Spanish American country except Paraguay. In 1962 taxes represented 9·8 per cent of GDP, rising to 11·1 per cent in 1966. At the same time, the composition of taxation became more regressive: taxes on income and property declined from 57·5 per cent of total central government revenues in 1962 to 39·1 per cent in 1966; taxes on domestic consumption rose during the same period from 6·3 per cent to 19·4 per cent of government revenues. A similar phenomenon occurred in Peru: between 1960 and 1966 the money value of receipts from direct taxes rose 55·2 per cent, whereas the money value of indirect taxes (especially sales taxes) rose 241·8 per cent.

It is a common assertion that 'there is likely to be a conflict

[5] M. J. Sternberg, *Chilean Land Tenure and Land Reform*, Ph.D. thesis, University of California, Berkeley, September 1962, Ch. IV.

[6] Banco de la Republica, *Cuentas Nacionales, 1962–1965.*

between rapid growth and an equitable distribution of income',[7] since a more equal distribution would lead to lower savings, fewer incentives and slower growth. The evidence from Spanish America does not support this view; indeed, the opposite could well be true. The lack of internal demand for agricultural and industrial products, the high propensity to import, capital flight, and the weak incentives to innovate and establish new activities could all be ameliorated if national income were redistributed in a more equitable fashion. Moreover, greater equality might lead to larger private savings (since the very rich probably save less than the middle classes), more productive investments (since the attractiveness of urban real estate would decline), an increase in productivity (as argued above), greater public savings (since tax evasion would diminish and the tax base would increase) and, if redistribution of income were accompanied by redistribution of land, higher agricultural output (since yields are greater on the minifundia than on the latifundia). Thus there need be no conflict between equity and growth and measures to redistribute income and wealth in the region should form an integral part of a strategy of economic development.

Growth of the Labour Force

The demographic pattern in Spanish America has two striking features, a high rate of growth of population and a high ratio of young dependents to wage earners. With respect to the former, the population appears to be increasing at an annual rate of about 2·6 per cent, although if Argentina and Uruguay are excluded the rate is nearer 3·0 per cent.

Table V:5

POPULATION GROWTH RATES
(per cent per annum)

Argentina	1·6
Bolivia	2·4
Chile	2·4
Colombia	3·1
Ecuador	3·4
Paraguay	2·6
Peru	3·1
Uruguay	1·4
Venezuela	3·4

Source: UN, *Monthly Bulletin of Statistics;* ECLA, *Statistical Bulletin for Latin America.*

[7] H. Johnson, *Money, Trade and Economic Growth*, Ch. VII, p. 153.

This 'population explosion' has made it much more difficult to achieve rapid increases in *per capita* income. If, for example, we assume the marginal capital-output ratio is 4:1 *and is unaffected by the rate of demographic increase*, a one per cent per annum increase in *per capita* output requires a savings ratio of 4 per cent, if the population is stable. If the population growth rate is one per cent, the savings ratio must be 8 per cent to achieve the same result. A 2·9 per cent rate of population increase would require 15·6 per cent of income saved to get a 1 per cent increase in *per capita* income. Thus the required savings ratio, under these assumptions, is the product of the marginal capital-output ratio and the sum of the desired rate of growth of *per capita* income and the rate of growth of population, i.e. $s = i/p\ (y + l)$.

Of course, it is unrealistic to assume that the capital-output ratio is unaffected by the rate of growth of the population. Indeed it is likely that the faster population increases the higher will be the capital-output ratio and, hence, the slower will be the increase in aggregate as well as *per capita* output. The reason for this is that a rising rate of population growth will lead to a change in the composition of investment from capital deepening to capital widening. That is, instead of raising '*i*' and thereby increasing '*p*', new investment will have to be spread over a large number of new entrants into the labour force in an attempt to prevent productivity from falling. As the population expands a large share of current capital formation will have to be devoted to providing basic facilities—water and sewerage works, houses, school buildings, roads, etc.—and the amount left over for investment in directly productive activities will diminish. Social overhead capital is generally characterized by long gestation periods and a low return on capital. Thus a shift in the pattern of investment in favour of social capital is almost certain to result in a higher agregate capital-output ratio, and consequently, a slower rate of growth of gross national product and *per capita* income.

The other feature of the demographic pattern also is of considerable interest. Chile's experience probably is fairly typical of Spanish America as a whole.[8] It seems, first, that the active population is declining relative to the total population. The groups at the extremi-

[8] See. B. Herrick, '*Efectos Economicos de los Cambios Demograficos en Chile, 1940–1960*', *Revista de Economia* (Santiago), Vol. 83–4, 1964. On Colombia see K. B. Griffin 'Coffee and the Economic Development of Colombia,' *Bulletin* of the Oxford University Institute of Economics and Statistics, May 1968, pp. 106–8; also see Chapter I, Part 1 above.

ties of the age spectrum are increasing relative to the middle group: the number of elderly persons is growing as life expectancy increases and the number of young people is increasing as the infant mortality rate declines. Secondly, the portion of the active population in the labour force is declining. Thirdly, the portion of the labour force that is actually employed is declining slightly, or at best is constant. Finally, the proportion of the employed labour force producing investment and consumer goods is declining, i.e. service industries are growing more rapidly than the production of goods. These final three problems are all related to a scarcity of employment opportunities.

A notable feature of the development landscape is that the actual labour force (whether 'employed' or not) is growing much less rapidly than the potential labour force. Leisure is being enforced on people because jobs simply are not available. The situation has become so severe that these young men and women of the 'leisured class' do not consider themselves to be members of the work-force at all and therefore do not consider themselves to be 'unemployed'.[9] Measured unemployment is only part of the problem however. Equally important are the scores of unutilized, partially utilized, and unutilizable labour existing in the economy. The economic and social cost of such wastage is beyond calculation. Rural 'disguised unemployment', overblown bureaucracies, swelling service industries, and the proliferation of petty retailers are symptoms of this social disease.

The published data on unemployment can be grossly misleading. For example, a survey of rural unemployment in the province of O'Higgins in Chile found not only that open unemployment in rural areas was extremely low but that seasonal unemployment appeared to be unusually low as well. The peak manpower needs during harvesting were met, to a large extent, by the temporary entrance into the labour force of members of the 'leisure class'. Thus many people who considered themselves members of the permanent work-force were enabled to work an average of ten or eleven months a year while non-members of the work-force had temporary employment for a month or so. Unemployment levels could persistently remain relatively low and thereby give a false impression of full employment.

[9] Thus in Chile, 93 per cent of the active population in the agricultural sector are males. On the other hand, those who migrate from rural to urban areas are predominantly females searching for employment. That is, unemployed females in rural areas are simply classified as economically inactive, becoming active only when they transfer to an urban environment. (See CIDA, *Chile: Tenencia de la Tierra y Desarrollo Socio-Economico del Sector Agricola*, pp. 14–15.)

But the economic fact is that when demand for labour rises the supply increases *pari passu*. The potential labour force in rural areas is much greater than the actual.

This reserve of potential manpower consists of unskilled, deskilled and unemployable workers. This 'reserve army of the unemployed' is not disguised, at least not in the sense in which Nurkse uses the term.[10] First, the reserve manpower is not readily transferable, particularly to other occupations requiring skills. Occupational mobility, in the sense of shifting from one type of skilled employment to another, is especially low among these people. Secondly, *industrial* labour is not in unlimited supply. Skilled labour is very scarce and very expensive. Finally, not all of the potential reserve of labour is available for full time employment; much of it is only seasonally available.

Thus the internal disequilibria represented by underutilization of the potential labour force must be attacked from at least four sides. Part of the general unemployment can be alleviated by increases in aggregate demand; if more work *opportunities* are provided, output will increase. Seasonal unemployment can be reduced by timed increases in specific demand, for example, construction of rural public works with seasonal labour.[11] The third step should consist of institutional reorganization and the retraining of labour which has lost its skills and become unemployable. Such institutional reorganization should include a land reform, which would have the advantage of eliminating the dependence of the rural worker on large landowners and bringing into cultivation previously unutilized land. A small portion of the underemployed workers would then have an additional factor of production with which to combine. Furthermore, it is necessary in many places to reorganize the composition of agricultural output so that seasonal unemployment can be reduced. This could be accomplished by diversifying crops in such a way that harvest periods come at different times of the year.

The injection of capital to increase productivity per man-hour should be the final stage. That is, the underdeveloped economies of Spanish America should first concentrate on raising the proportion of the population that is in the actual employed labour force. Then the proportion of working hours per man should be increased so that all partially employed workers become fully employed, and the entire workforce is working a full day, five or six days a week. It is

[10] See *Problems of Capital Formation in Underdeveloped Countries.*

[11] See T. Balogh, 'Agriculture and Economic Development', *Oxford Economic Papers*, February 1961.

time enough to start injecting capital, and raising the capital-labour ratio, once these tasks have been completed.

Capital Intensity

Frequently it is observed that in the less developed countries the techniques of production fail to reflect the prevailing factor proportions; that is, the employed techniques are too capital intensive. Four reasons may account for the excessive capital intensity of the methods of production. First, the equipment made by mass-production methods in the advanced economies is likely to be relatively cheap.[12] Second, the market prices for factors of production may not reflect the 'intrinsic value' or 'opportunity cost' of the factors.[13] Market wages may be considerably higher than the real cost of labour to the nation. This will be true if unions can introduce a monopoly price of labour, or if welfare considerations have dictated a minimum wage in industry, or if social security payments are unusually high, or if the presence of large-scale rural unemployment means that the marginal productivity of labour in agriculture is near zero. In addition, interest rates often fail to reflect the true scarcity of capital. International lending institutions usually charge moderate rates of interest, and there is a tendency for less developed countries to organize cheap credit facilities for certain types of small enterprise. These two factors make it probable that interest rates on the whole will have a downward bias. Furthermore, even a moderate rate of inflation will greatly reduce the *real* rate of interest. For example, if the nominal rate of interest is 17 per cent and prices are rising at an annual rate of 10 per cent, the real rate of interest is only 6·3 per cent.[14]

Discontinuities in the production function occasionally are cited as a third reason to account for 'structual disequilibrium at the factor level'. The importance of this argument easily can be exaggerated. It is possible that some factor coefficients are fixed in *production* but it is doubtful that they are fixed for the *production unit*, i.e. the firm. For example, it is quite likely that production techniques for steel are fixed and discontinuous, but it is doubtful that there are rigid coefficients in the supervisory, administrative, clerical and main-

[12] R. Nurkse, *op. cit.*, p. 45.

[13] J. Tinbergen, *The Design of Development*, p. 39.

[14] This was calculated using the formula $r = \frac{(1 + n)}{(1 + v)} - 1$ where r = the real rate of interest, n = the nominal rate interest; and v = the percentage rate of price increase.

tenance departments of steel companies. Fourth, as discussed in Chapter II, there may be a 'demonstration effect' on the side of production as the poorer countries try to imitate the production methods of the wealthy nations.

For these reasons it is quite likely that production techniques in Spanish America are more capital intensive than one might expect from merely observing the relative prices of 'capital' and 'labour'. This is particularly true on the margin, i.e. for new investment.

The Macro-economics of Inflation and Unemployment

The four structural elements of our economy now may be brought together to give us a dual condition for long-run equilibrium growth. The first condition states that the additions to the labour force must be equipped with capital at the ratio of capital to labour appropriate to existing production techniques as determined by technical knowledge and the relative real scarcity of labour and capital. The second condition for stable growth is that the net investment thus required by the growth of the labour force must equal net saving at full capacity output. That is, aggregate demand must equal full employment aggregate supply.[15]

The dual requirements for equilibrium growth can be put in the form of a simple equation:

$$ps = il.$$

It should be evident from our discussion on the previous pages that the equilibrium condition does not hold in most of the Spanish American countries. Productivity and the savings ratio are low; methods of production are relatively capital intensive. The population growth rate is rapid and consequently the rate of growth of the potential labour force is high. The implications of the disequilibrium can be shown in a four-quadrant diagram.

Given the above observations about the *relative* values of our structural elements the disequilibrium becomes clear. If we start with a potentially available labour force of L_a and know its average productivity (p) we can determine aggregate supply (Y_s). That is, $Y_s = pLa$. Starting at L_a and going the other way around the diagram, we can see that the increase in the labour force will be z. In order to equip all these workers at the existing capital-labour ratio,

[15] The analysis of this section relies heavily on a version of the Harrod-Domar growth model developed by John Power. See J. Power, 'Economic Framework of a Theory of Growth', *Economic Journal*, March 1958; 'Capital Intensity and Economic Growth', *American Economic Review* May 1955.

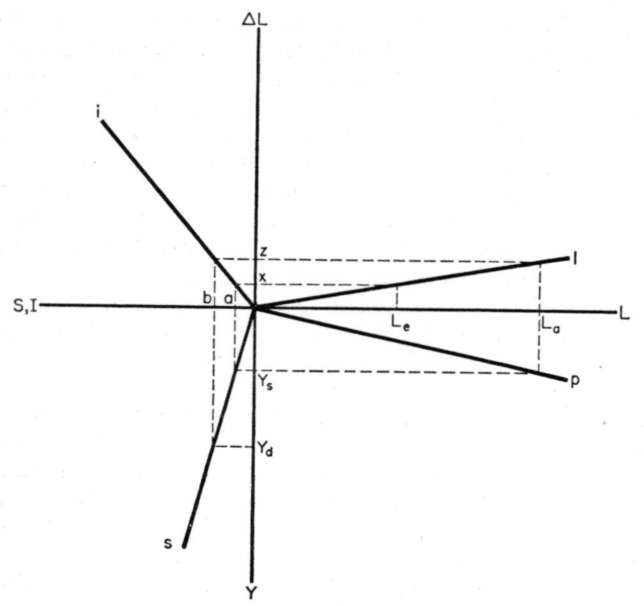

Figure V:1

b amount of investment will be required. Investment of b, times the multiplier (l/s) will generate aggregate demand of Y_d. That is, $Y_d = L_a.l.i/s$. Thus with an available labour supply of L_a the total supply of net output will be Y_s. At the same time the total demand for net output will be Y_d. As $Y_d > Y_s$ there will be inflation. Similarly, the employed labour force will be only L_e.[16] As $L_a > L_e$ there will be unemployment of one sort or another. In other words, our Spanish American economy will experience *both* inflation and unemployment. Just how much inflation or how much unemployment will be suffered is not immediately obvious from the diagram because the model does not give a determinate solution. As the diagram is drawn, the economy may have *either* unemployment equal to $L_a - L_e$ *or* inflation equal to $Y_d - Y_s$ *or*, what is most likely, a combination of inflation and unemployment.

What has happened is that investment equal to full employment saving is not sufficient to equip the growing labour force at the

[16]An aggregate supply of Y_s will only generate enough savings to finance 'a' amount of investment and equip 'x' amount of the increase in the labour force. This economic activity is sufficient to employ only L_e workers.

selected capital-labour ratio, so that some of the workers remain unequipped. There is redundant labour but the existing capital stock is fully employed. To reach an equilibrium growth path it will be necessary to increase the savings ratio and raise productivity or reduce the rate of growth of the population and adjust the capital intensity of the methods of production. None of these solutions may be easy. It is important to notice, however, that the same policies which are needed to combat inflation and unemployment will serve also to raise the rate of growth. In these terms there is *no* incompatibility between growth and monetary stability.

A glance back to the equation on page 177 of this chapter will enable one to see clearly that the *per capita* rate of growth will increase if s is increased, p is increased, i decreases, and *l* decreases. These parameters, however, may interact in a rather complicated way. For example, an increase in s will raise the rate of growth of *per capita* income. A higher growth rate may encourage a faster rate of population growth (*l*) and this in turn may stimulate investment in housing, thus raising i. A shift in the composition of investment towards housing will raise the capital output ratio, i/p, and thereby reduce somewhat the rate of growth of national income.

On the other hand, using capital intensive techniques of production will raise i; however, over a certain range p may rise equally as much or more. Assuming consumption can be held at some sort of a subsistence level, labour surplus or saving—the difference between the productivity of labour and consumption—will increase. In that case, raising i may have no ultimate effect on the capital-output ratio but instead will increase s and consequently the rate of growth of national income.

What the analysis so far has shown is that the *mere manipulation of monetary policy cannot eliminate the structural disequilibrium.* Monetary policy can only influence the *form* in which the disequilibrium appears. If the supply of money is drastically reduced the disequilibrium will appear as an employment gap and low growth rate.[17] If liberal credit policies are pursued, unemployment will diminish and the disequilibrium will appear as an inflationary gap. Many governments follow a compromise in their monetary policy

[17] The 'monetarists' frequently claim that inflation is caused by a lack of financial 'discipline' and large government deficits. They fail to notice, however, that a reduction in the annual rate of inflation (achieved, say, by a contraction of the money supply) will tend to increase the government's deficit or reduce public savings. See the appendix to this chapter for a demonstration of this proposition.

and have both inflation and unemployment. Others follow a peculiar type of 'stop-go' policy and oscillate wildly between stagnation and inflation. Thus how any particular economy behaves will depend (a) on the value of its structural elements and (b) on the policies pursued by the government.

In summary, we have found that—except where exports are expanding rapidly—the structural elements of the Spanish American economies combine to produce a growth disequilibrium. This disequilibrium may be written in the form $sp < il$, but it is likely to manifest itself in an economy in the form of a simultaneous presence of inflation and unemployment. Policies which are likely to promote growth also are likely to ameliorate the inflation. The best and most straight-forward policy would be to increase savings (and investment).

2 SECTORAL BOTTLENECKS

One cannot fully understand the process of inflation in Spanish America without disaggregating the economy and considering a few crucial sectors where specific bottlenecks may arise. From our previous discussion we know that inflation in Spanish America is not primarily a monetary phenomenon. It is caused by structural disequilibria in the nation's economy and can be controlled only by ensuring compatibility among the various elements of national expenditure. It may clarify our thinking if we define total national expenditure or ouput (Y) as the sum of expenditures on essential consumption items (E), non-essential or luxury consumption goods (N), private investment (Ip) and state investment (Is).

Perhaps a word needs to be said about definitions. Just as the division of expenditure between consumption and investment is in some sense arbitrary so is the division of consumption into E and N goods in some sense arbitrary. Nevertheless, it is of considerable analytical importance. The definition of 'essential consumption' can be no more precise than the use to which it is put. In all cases it must be interpreted in a pragmatic, commonsense way. Two points of clarification, however, may be useful. First, essential consumption does *not* refer to that minimum necessary to maintain life; it is not a physical subsistence minimum. Second, to the extent that it corresponds to an income floor, it corresponds to a *culturally determined* consumption minimum; this will be higher than a bare life-maintaining level of consumption, and will vary from one country to another, as for example from Ecuador to Uruguay. That is, our notion of a subsistence level is quite similar to what Ricardo called 'the natural price of labour'. Similarly, by 'investment' is meant socially productive net capital formation. By this definition luxury housing in Venezuela and automobile factories in Argentina are examples of socially unproductive investment. From the point of view of the growth of Spanish America and the welfare of the majority of its citizens they are examples of luxury consumption and are so treated below.

Thus we have an identity: $Y = E + N + Ip + Is$. The surplus of production (Sr) consists of expenditure on all goods other than necessities—$Sr = Y - E$—and savings (S) represent that part of the surplus which is used for socially productive investment: $S = Sr - N$.

Growth proceeds by restraining non-essential consumption or by maximizing the growth impact of expenditures out of surplus. Essentially this means maximizing the ratio S/Sr although we must not forget that certain types of consumption expenditure are complementary to capital formation. Thus the crucial task of development policy consists in finding an optimum combination of private investment, growth-inducing consumption, and public capital formation. Vigorous pursuit of this goal is likely to generate and require consistent inflationary pressure. The inflationary pressure can be controlled and rapid price increases prevented by ensuring appropriate relationships between E and Y; N and S_r; I_p and I_s.

First, the supply of essential consumption goods must be in the proper relation to the given level of disposable income. This means essentially that food production must keep pace with population increase and the rise in *per capita* income. Second, expenditure on luxury items must be sufficiently restrained to provide out of surplus adequate savings to finance both private and state investment. Finally, private investment must be sufficiently restrained so that total investment does not consistently exceed planned savings, i.e. adequate savings must be available to finance planned public investment. Inflation will result if either desired consumption of necessaries is greater than the supply of these goods, or if non-essential consumption out of surplus plus total investment expenditure is greater than actual surplus, or if desired private investment is greater than the share of savings made available to it by private savers and the public authorities. Price increases can be restrained best by controlling each of these expenditure categories; general monetary measures will not be efficient.

Expenditure unbalance and inflation may arise from any sector. For example, if food output fails to increase in step with additional demand while the population grows and industrial production increases, agricultural prices—in the absence of effective controls—will rise.[18] The higher urban incomes, resulting from the greater output of industry, will soon bid up the price of food. The higher food prices will increase the cost of living and wages will be forced up. A spiral, alternating between higher food costs, higher wages, and higher prices in general, could easily begin now. Whether rising agricultural prices will be effective in stimulating additional food and fibre output, and thereby dampen the inflation, will depend upon particular tenure institutions, distribution mechanisms, knowledge and incentives.

In Chile and Argentina, for example, these elements seem to have combined to limit the expansion of output. In Chile since 1940 and in Argentina since 1949 agricultural output has lagged behind the rise in demand. At the same time the prices of agricultural products have risen relative to industrial prices, general wholesale prices and relative to the prices of agricultural inputs.[19] It would appear from this evidence, therefore, that structural rather than pricing difficulties are the major cause of agricultural stagnation, although it is possible to argue that agricultural supplies have not increased rapidly because their prices still have not risen enough. In any case, the key to solving this aspect of the inflation problem lies not in reducing aggregate demand but in increasing supply from a specific sector.

We have selected agriculture to illustrate our thesis that inflation may arise not only from a general excess of aggregate demand but also from sectoral supply deficiencies because poor performance in the agricultural sector is a feature common to most Spanish American economies. As can be seen in Table V:6, in most cases agricultural output has failed even to keep up with population increase, let alone to satisfy the additional demand created by rising *per capita* incomes.

An equally relevant example arises from the policy of import substitution pursued by the various countries of the region. Many of the newly established industries in Spanish America were created, or at

[18] In theory, it is possible that industrial prices could fall, but the downward price rigidity in industry suggests that this is not likely. Alternatively, agricultural prices could remain unchanged if food could be imported in unlimited amounts. In view of the slowly growing export earnings and the protectionist policies of the Spanish American governments this possibility, too, is unlikely.

[19] See E. Eshag and R. Thorp, 'Economic and Social Consequences of Orthodox Economic Policies in Argentina in the Post-War Years', *Bulletin* of the Oxford University Institute of Economics and Statistics, February 1965; O. Sunkel, '*La inflacion Chilena: un enfoque heterodoxo*', *Trimestre Economico*, October–December 1958; translated and reprinted in *International Economic Papers*, No. 10.

Table V:6

PER CAPITA AGRICULTURAL PRODUCTION
(1957–59 = 100)

	1965	*1966*
Argentina	94	99
Bolivia	95	95
Chile	96	92
Colombia	97	98
Ecuador	100	98
Paraguay	100	93
Peru	103	99
Uruguay	108	106
Venezuela	121	118

Source: US Department of Agriculture.

least strongly encouraged, by protectionist policies. Yet the limited size of the domestic market permits the efficient operation of only one or two firms in each industry; consequently oligopolies, with their traditional disregard for price competition, flourish. Indeed, in some countries, e.g. Chile,[20] industrial concentration may be rising. In addition, trade unions attempt to exert a strong upward pressure on costs. Higher costs soon lead to higher prices. The occasion for further price rises is provided by the mechanism of collective bargaining when some wage agreements exceed the average and create a problem of wage differentials. Thus, in the absence of specific controls a cost-push inflation will result from the combination of business oligopolies, strong labour unions and the inevitable spread-effects.

The 'wage-problem', however, is much more than the manifestation of particular institutional arrangements. It reflects the social dynamics of a vociferous and important group in the society. It is evident to even the most casual observer that the working population of Spanish America is demanding higher real incomes. These demands can be satisfied in only two ways: either real wages must be increased from more rapid increases in productivity and the rate of growth of output, or the share of wages in the national income will have to rise. These demands for higher real incomes from a large and occasionally well-organized segment of the population will have to be taken as 'given' in any equations concerning inflation and exchange rate policy. The particular institutions and bargaining

[20] See Instituto de Economia, *Desarrollo Economico de Chile, 1940–1956*, Ch. VIII.

mechanisms employed are only reflections of the social attitudes which underlie them; it is the *attitudes* which must be taken into account, not just the institutions. It will not be possible to 'trick' the workers into accepting a reduction in wages, and schemes which, for example, propose to give a *reajuste* or wage readjustment one day and devalue the next are doomed to failure. Aside from the social tension and heightened class conflict such an approach would entail—effects which are undesirable in themselves—it simply would not work. In the long-run the only people who would be fooled would be the naïve policy-makers. The region's problems of slow growth, inflation, and persistent trade deficits all stem from the same source—an insufficiency of savings. This is the fundamental problem and policies which do not attack this are, by comparison, mere 'gimmicks'.

An increase in savings, a moderation of the wage spiral and a rise in productive investment cannot be achieved without the co-operation of organized labour. Unfortunately, few Spanish American governments enjoy the confidence of the union leaders and the working classes. The mass of the population has come to believe that its aspirations and objectives are not shared by the government, i.e. by the groups which monopolize effective political power. There is thus a constant clash between the unions (or, more generally, the urban and rural workers) and the government. This clash has its roots in the social and political structure of the society. Arthur Lewis is one of the few non-Latin American economists who has recognized this. He says, 'Of all social classes the most reactionary social class is the class of great landowners. . . . Nowadays the power of great landowners has been broken all over the world except in the Middle East and in Latin America, and so Latin America is the most politically reactionary of the continents. It is, for this reason, the angriest continent, and therefore about the last place where the trade unions are likely to agree to any kind of wage control. In this sense the wage spiral in Latin America is fundamentally political, and cannot be eliminated without fundamental political change.'[21]

The Case for Inflationary Finance

One must not assume automatically that inflation is undesirable since a case can be made for it on both historical and theoretical grounds. According to Rostow, 'In Britain of the late 1790s, the United States

[21] W. Arthur Lewis, 'Closing Remarks', a paper presented to the Conference on Inflation and Growth in Latin America, Rio de Janeiro, January 3–11, 1963, pp. 7–8 (mimeo.).

of the 1850s, Japan of the 1870s, there is no doubt that capital formation was aided by price inflation. . . .'[22] How is this possible?

As output increases over time it will be essential to expand the money supply in order to avoid deflation. First, as national income increases the need for money for transaction purposes will increase. Second, as the economy develops, the non-monetized sector will decrease and the size of the monetized sector will increase; this will require an additional expansion of the money supply. Third, as incomes rise there appears to be a tendency for cash balances to increase faster than the rate of increase of real *per capita* income, i.e. the money-income ratio rises.[23] The combination of these three factors is not likely to result in a significant expansion of credit, yet any further increase in the money supply may be inflationary. Whether the inflation is beneficial to the economy depends (a) on whether it is a profit inflation or a cost-of-living inflation and (b) if a profit inflation, whether the marginal propensity to save out of profits is high.

Ideally an inflation could have two beneficial effects. First, rapid price increases might be successful in inducing the entry of non-wage labour into the market economy. If this should occur the price rise would be partially self-correcting and would have the advantage of increasing output by increasing effective factor inputs.[24] This effect is likely to be small and, in fact, it was completely ignored in Part 1 when we concluded that unemployment and inflation are largely alternatives which have little influence on real output.

The second effect depends upon a redistribution of income.[25] Unemployed labour from rural areas could be used in capital formation and the workers could be paid through credit creation. This new money would generate inflation, but since the output of essential consumption items is being kept constant the inflation will (i) serve to redistribute *E* goods to the newly employed and (ii) raise the income of agriculturalists and industrialists. If these two groups have a higher marginal propensity to save than the other groups in the

[22] W. W. Rostow, 'The Take-Off into Self-Sustained Growth', *Economic Journal*, March 1956.

[23] See M. Freidman, 'The Demand for Money: Some Theoretical and Empirical Results', *Journal of Political Economy*, August 1959.

[24] See S. P. Schatz, 'Inflation in Underdeveloped Areas: A Theoretical Analysis', *American Economic Review*, September 1957.

[25] See W. A. Lewis, 'Economic Development with Unlimited Supplies of Labour', *Manchester School*, May 1954.

nation, effective savings will have been increased and capital formation will have been accelerated. This process is known as 'forced savings'.

Inflation continues until the additional savings from newly created profits equal the amount of additional investment. Thus the ratio of profits to national income must increase. If the profit receivers spend their additional income on investment items the inflation will continue. Eventually, however, technical progress plus the output from the newly created capital will increase supply enough to damp the price rise; inflation for purposes of financing capital formation is in this sense 'self-destructive'. Clearly inflation is but an inefficient alternative to taxation. Furthermore, it requires a large industrialist class with a high marginal propensity to save in order to be effective at all. Yet, if a large increase in output quickly follows the new capital formation, it can be (and has been) a useful tool in accelerating economic development.

The Case against Inflation

There are, however, certain obvious disadvantages which are associated with an inflationary policy. First, inflation may lead to a regressive redistribution of income; the rich get richer and the poor get poorer. This in turn may cause severe social problems. Second, it may dull incentives to efficiency because inflation, if it is strong enough, will bring profits to all entrepreneurs, even the inefficient ones. It usually is argued that the creation of social tension and the encouragement of inefficiency depend upon the ability of inflation to alter the distribution of income. The experience of Spanish America has shown, however, that those groups whose incomes would normally be adversely effected by inflation have been quite successful in building defences against persistent price increases. That is, the redistribution problem is associated not so much with a high *rate* of inflation as with *changes* in the rate. A *constant rate* of price increase can be defended against eventually, but it is much more difficult to neutralize the effects of periodic *changes in the rate* of price increase. Social tension (strikes, demonstrations) and inefficiency may persist, not because income has been redistributed, but precisely because a great deal of time and energy are devoted to ensuring that in the long run income is *not* redistributed. In other words, once people come to expect continued inflation the redistribution mechanism fails to operate, savings do not increase and the advantages of an inflationary policy disappear; at the same time the disadvantages of a redistribution may remain because the *threat* of falling incomes will elicit contervailing action whose secondary effects are similar and adverse.[26]

Equally important are the effects rapid price increases may have on the allocation of investment. Enterprises whose prices are subject to regulation and are not easily raised, e.g. utilities, or firms which may obtain capital through the issue of bonds, e.g. railroads, may find it increasingly difficult to obtain finance. Social overhead capital may be a bottleneck to growth and it might seriously hamper the development effort to discourage investments in this area. Certainly the disinvestment in the nationalized railroads in Chile, for example, and the (until recently) extremely slow expansion of telephone communications have not helped economic development in that country.

Inflation, when combined with restrictions on the importation of luxuries, will lead to attempts by domestic investors to replace or provide substitutes for the restricted imports. A danger will arise that the less 'essential' is a good (and hence the more difficult it is to import it) the more profitable it will be to produce it domestically. A tariff policy which is used to alter the composition of imports from consumption items to capital goods does not *ipso facto* increase the volume of savings. Such a policy may induce an efficient use of scarce foreign exchange, but such a policy *by itself* will have no effect on savings or the rate of growth. Savings will not increase unless the income that would have been spent on imports is saved instead of being spent on domestically produced consumer goods. The *composition* of imports and the *composition* of investment will be changed but the *level* of both will remain unchanged. Investment in luxury industries represents a poor use of scarce capital, and a good case can be made for government controls in order to prevent such a use of resources.

Perhaps even more important, luxury urban construction and investment in inventory holdings are stimulated by inflation. Not only is the proportion of construction to total investment increased, but the composition of construction activity appears to shift away from industrial buildings to commercial and luxury housing. These changes in the level and composition of the building industry reduce the aggregate productivity of capital and slow down the rate of

[26] An unsolved question is why the price system does not explode into hyperinflation. One of the reasons is that two important price groups—wages and the administered prices of public utilities—lag behind all the others; this dampens the price increases. Another reason seems to be simply that Spanish Americans have learned to live with extraordinarily high rates of inflation without completely losing confidence in the monetary system. This 'learning through living' effect appears to be the more important of the two explanations.

growth. Furthermore, the composition of *total output* is changed; production is switched from perishables to durables, i.e. to those goods in which it is possible to realize capital gains. The tendency to speculate increases.

We have argued that low real rates of interest may influence the choice of technique. Some economists also argue that inflation, by reducing the real rate of interest, acts as a subsidy to capital and especially to capital intensive goods. Hence, those goods which are more capital intensive than others will be encouraged and in this way the composition of output will be influenced. Our own conclusion is that this effect is likely to be small. Capital intensive industries are frequently social overhead capital and monopolies with government controlled prices. These prices are kept low during an inflation and investment in these industries is discouraged, thereby creating major problems for development. Secondly, inflation has led to speculation and to the production of *durables* as oposed to perishables, whether or not their production is capital intensive. Similarly, inflation has led to investment in real estate and construction because these are areas where capital gains are available. Thus it seems likely that low real rates of interest influence the *way* things are produced but not *what* is produced.

The part played by services in national output has been rising in most Spanish American countries. It is not clear to what extent this tendency is due to inflationary forces. The shift from goods to services must be due in large part to (a) the increase in commerce, (b) the distribution of income, (c) the differences between the sectoral income elasticities of demand, and especially (d) the limited availability of employment opportunities in industry. That is, the shift in the composition of output reflects both supply and demand factors. Some of the increase in commercial activity can be attributed to inflation. It is also clear that the presence of a considerable volume of unemployment may force aspiring workers to enter service industries and become, in effect, disguised unemployment.[27] It is less clear that the inflation is responsible for unemployment. If our analysis in Part 1 is correct, inflation and unemployment are largely alternatives; on this basis the shift of employment to service trades cannot be attributed to persistent price increases. On the other hand, the change in the composition of investment from manufacturing to construction activities will reduce employment opportunities in industry and thereby force labourers into the services sector. Inflation

[27] This is true in part because minimum wage legislation does not apply or is not enforced in many service industries.

is responsible for this effect, and thus may be partly responsible for the general tendency of workers to shift to services. This takes care of the supply side. On the demand side it seems evident that services are largely consumed by the wealthy. Thus an unequal distribution of income will increase the demand for services. Inflation, however, has not been remarkably successful in redistributing income from the poor to the rich; many groups by now have been able to build defences against price increases. The extremely skewed distribution of income is due to other causes; thus inflation cannot be charged with increasing the *demand* for services. The issue, therefore, remains in doubt; on balance one might conclude that inflationary forces working on the supply side may have increased slightly an autonomous tendency for the production of services to increase relative to the production of goods.

The usefulness of inflation in financing capital formation rests on its ability to shift the distribution of income toward profit takers with high savings propensities. It is expected that profit earners will save some of their increased income. Other groups, particularly those on fixed or semi-fixed incomes, will suffer a decline in their real income. In an attempt to maintain their customary standard of living these groups actually may dis-save. If this occurs, a large shift in income to profit receivers may be necessary to effect even a very small net increase in savings. It seems likely, however, that the individuals who will be adversely affected by inflation will be people with little previous savings or ability to obtain credit, and consequently they will be *unable* to dis-save. A danger remains, however, that an inflation, which is due to a lack of savings, may leave the distribution of income substantially unchanged and at the same time destroy what savings habits exist. Far from accelerating growth, inflation *may* retard it.

Rising domestic prices can have extremely damaging effects on the foreign exchange position of a country. As internal prices rise, the price of imports will decline relative to the price of domestically produced goods, and the volume of imports therefore will rise. In addition, the high internal demand for goods accompanying inflation increases the danger that goods formerly exported will be consumed. Finally, as inflation erodes the purchasing power value of paper assets an outward capital flight may occur. The latter has been an important savings leakage in nearly all the Spanish American countries. Import restrictions, export subsidies and capital controls become essential. Eventually currency devaluation becomes necessary. The devaluation may have three unfortunate effects. First, by raising the internal price of both imports and exports it will lead to increased inflation. Second, if exports expand sufficiently to lower the world price of primary products, it may lead to a decline in the terms of trade. Third, by lowering the dollar costs of the foreign owned companies which produce for export, the profits of these corporations will increase and, unless tax rates are properly adjusted, capital repatriation abroad will accelerate. Thus devaluation may be a costly policy.

The Administrative Problems of a Policy of Inflation

We have seen that if wages lag behind prices inflation will be successful in redistributing income towards the profit sector. If, furthermore, manufacturers have a high savings propensity inflation can successfully finance capital formation. Not only will total surplus be increased but the proportion of surplus productively invested will increase.

Inflationary finance can be effective *only* if the *government administration is efficient and impeccably honest.* Strict controls will have to be operated, especially over foreign exchange. It is important that the tax system be constructed so that marginal tax rates are higher than average tax rates, for only then will government expenditures ever reach a point where they can be maintained without further creations of money. If, as is usually the case in Spanish America, marginal tax rates are lower than the average rates, inflation will lead to a large government deficit and this, in turn, will accentuate the inflationary process. Price increases must be moderate, coming in spurts and alternating with periods of stability; otherwise people will form expectations about continuing price increases which will neutralize the distributive mechanism. Projects to be financed by credit creation must be judiciously selected so that large and prompt increases in output follow the capital formation. Since in many Spanish American countries the oligarchy has direct access to credit through family owned banks and insurance companies, firm control over the banking system—and perhaps nationalization—will be required.

If labour unions and business oligopolies form an important element in the economy, so that it is difficult to transfer income to savers, it is harder to operate a profit inflation efficiently; but it can still be done. The government will be forced to manage a disequilibrium system. Foreign exchange controls and tax policy will have to be supplemented by rigorous price controls, rationing of essential consumer goods, and allocation of investment by licensing and priority schemes. The economy now will be virtually fully planned and managed. Thus under 'optimum' planning conditions a large class of private industrialists and government entrepreneurs with a high marginal propensity to save will enjoy rising industrial prices and profits while agricultural prices decline as a result of increases in productivity. These measures, combined with partial tax exemption on savings used for productive investment, might go part way in raising the savings ratio and accelerating the rate of capital formation.

To the extent therefore that (a) inflation redistributes income to profit takers and (b) the propensity to save out of profits is relatively high, inflationary financing can accelerate development. There are two important points here. First, it is difficult for the government in many Spanish American economies to raise sufficient funds through taxation to finance the desired amount of state investment: tax evasion is widespread. Inflation in this case can act as a subsitute for taxation. Second, and more important, inflation may raise profit expectations.

A major problem in development policy is to ensure that private investment fulfils the role allocated to it by the government planners. Too often no effort is made to see that the planned target for private investment actually is fulfilled; frequently it is just a number selected so that the sum of state and private investment equals the aggregate investment target required to achieve the projected rate of growth. Ideally, inflationary tension, by shifting income to profits and by creating expectations of a further increase in profits, provides the means for and the encouragement to private investment. The private sector, far from being a drag on capital formation, becomes a positive stimulus to economic development.

Aggressive industrialists are then likely to boost private investment above the planned rate. There is the danger that the government, in order to prevent an excessive inflationary burst, will cut back on state investment. This is where controls come in. Once businessmen are eager to invest in a variety of directions planning in the private sector becomes simple: all the government has to do is keep the lid on. Planning in this sector then will consist essentially of the negative act of preventing capital formation in undesired areas. Such a mechanism appears to have worked for a while in India, a country with a civil service better than the average. Inflationary pressure kept profits and private investment in industry bouyant. Government controls served the dual function of regulating private capital formation and keeping the inflation within moderate bounds.

A large selection of measures is needed by the government in order to perform the functions of guiding investment into the appropriate channels. The most useful devices will consist of *selective* financial and fiscal measures plus a full complement of *direct* controls. For example, import restrictions will be necessary to prevent a deterioration of the balance of payments. In so far as domestic investment requires the importation of some capital equipment, import restrictions will also be useful in controlling private capital formation. On the other hand, licensing schemes may be needed to prevent luxury urban construction. The conclusion that emerges is that general monetary policies are less likely to be useful than other, more discriminatory, measures.

It is obvious that the optimum conditions for running an inflationary policy of this sort do not exist in Spanish America. For the most part inflation has been unable to increase much further the already extremely unequal distribution of income. The savings propensity of the rich is low. The government bureaucracy is inefficient. The tax system is poor. Price increases have been immoderate and persistent. People have become very sensitive to price changes and price expectations are highly elastic. On all counts, therefore, these governments are unable to operate an efficient inflationary policy. It would appear, then, that the time has come to stop trying to live with unplanned inflation and start planning to live without inflation.

The Stabilization Policies of International Institutions

So far three problems have been discussed in this chapter: the cause of inflation in Spanish America, the need for and desirability of its restraint and, briefly, the method of its control. It has been argued that structural imbalances are the primary cause of inflation in these economies. Thus it is *existing* structural distortions which create *particular* supply bottlenecks and savings scarcity; these bottlenecks, in turn, generate inflation, and the inflation creates further problems. Thus the 'cause' of inflation is due as much to deficient supply or lack of savings as to excess demand. This view, of course, is not accepted by everyone.

The International Monetary Fund, for example, seems consistently to take the view that the crucial structural disequilibria are the *consequence* and not the cause of inflation.[28] Furthermore, it is implied that if the symptoms of inflation were removed the rate of growth of output would increase. It is on this basis that the IMF usually advocates monetary retrenchment. One should note that this

[28] For a clear statement of the IMF position see *A Report on the Process of Inflation in Chile Together with Recommendations*, 1950.

is a simple neo-classical analysis which uses all the neo-classical assumptions. In particular, it is assumed that the economy is not fragmented or poorly articulated, so that *general* monetary measures —believed to have no directional effects—taken against a *general* excess of demand will be necessary and sufficient.

First, of course, it simply is not true that monetary policy has no directional bias. Post-war experience in Britain and the United States has shown that monetary policy is slow to work, that the public sector does not respond to it and that the effects of a general monetary policy are unequally allocated among various groups. It would be purely fortuitous if the groups or sectors that received the brunt of the impact were those groups which needed most to be restrained. Secondly, the analysis in this chapter seems to indicate that excess demand may not be widely diffused but may be located in particular sectors. It therefore seems likely that general monetary measures, sufficiently strong to prevent sectoral inflation, are likely to impart a deflationary bias to the entire economy and thereby hamper growth. The elimination of inflation by monetary retrenchment alone, far from encouraging growth, will seriously retard it.

The application in Argentina during the period 1958–63 of policies advocated by the IMF led to a fall in *per capita* consumption of about 20 per cent and no permanent decline in the rate of price increase.[29] The success achieved in Peru since 1959, based largely on a rapid and autonomous rise in exports, 'has been in spite of, not because of, her attempts at stabilization'.[30]

The experience in Chile with the stabilization programme in the last half of the 1950s is equally instructive. Beginning in 1956 a severe credit restriction was applied, investment expenditure was reduced, the automatic wage increase to compensate for inflation (*reajuste*) was abandoned and some fifty commodities which had heavy weights in the formation of the consumer price index were subjected to price controls. For a while the rate of price increase was reduced substantially. The 75 per cent increase in prices during 1954–55 was reduced to as little as 7·7 per cent in 1961. From then onwards prices began to rise again at an accelerated rate and during the twelve month period ending July 1963 prices increased 48·8 per cent.

Thus the stabilization programme on its own terms was only partially successful. The rate of price increase was only reduced tem-

[29] See E. Eshag and R. Thorp, *op. cit.*

[30] R. Thorp, 'Inflation and Orthodox Economic Policy in Peru', *Bulletin* of the Oxford University Institute of Economics and Statistics, August 1967, p. 205.

porarily and it never was stopped completely. At the same time other indicators showed that the decline in investment associated with unfavourable expectations and monetary retrenchment seriously affected the performance of the economy. Unemployment in Santiago rose sharply and continued to rise until the devaluation of January 1959 had time to take effect. (In March 1959 the unemployment rate in Santiago was 10·4 per cent.) The rate of growth of output similarly was affected. Growth was very slow in Chile throughout most of the 1950s. From 1954 through 1959 the *per capita* gross domestic product expanded at an annual rate of minus 0·53 per cent. Finally, it appears that the distribution of income became somewhat more unequal during this period. In no sense, therefore, can the Chilean stabilization programme be deemed a success. The majority of the working population was made worse off in three ways: through an increase in unemployment, through a negative rate of increase of *per capita* income and through a regressive redistribution of income. Economically and socially, a more unfortunate policy is hard to imagine.

Throughout Spanish America the single most important effect of the 'stabilization' policies pursued has been a decline in investment activity and employment. Instead of reducing aggregate demand *relative* to supply, severe credit restrictions and the accompanying measures have done little more than reduce demand *and* supply. In this way the growth of output and *per capita* income is severely damaged.

At times it seems that the IMF almost welcomes this consequence of its policy. Professor Hirschman quotes the Fund as saying,

'A credit restriction to be effective must force businessmen to sell goods at prices lower than they had anticipated, oftentimes at a loss; it must make it financially impossible for them to increase wage rates, and it must cause a certain minimum amount of unemployment.[31]

An economist working for the World Bank informs us that his organization occasionally co-operates with the IMF by reducing development loans in order to 'fight' inflation.[32] In other words, the combined policies of these two important international institutions seems to be to combat inflation by creating unemployment and reducing investment. By concentrating exclusively on demand

[31] A. O. Hirschman, *Journeys Towards Progress*, p. 233; quoting the 1950 report of the IMF to Chile.

[32] R. E. Carlson, 'The Role of the World Bank in Latin America', in P. D. Zook, ed., *Economic Development and International Trade*.

factors, and ignoring policies which tend to increase supply, these two organizations have hampered the development effort of the Spanish American economies and, at the same time, they have been largely unsuccessful in eliminating inflation. A greater policy failure is difficult to contemplate.

3 DEVALUATION AND INTERNAL STABILITY

As we have seen, a major cause of Spanish America's chronic balance of payments problem is the persistent increase in domestic prices. The International Monetary Fund's position with respect to inflation has been to encourage credit contraction. Similarly, the Fund's position with respect to the payments deficit has been to encourage deflation and, when necessary, devaluation. In the remaining portion of this chapter we will argue that this apparently simple solution frequently is inappropriate in the development context of Spanish America. We then will go on to suggest a possible mix which in some circumstances may help to control the associated problems of inflation and the balance of payments.

The Meaning of an Equilibrium Exchange Rate

According to the Bretton Woods system a country should devalue only when its balance of payments is in 'fundamental disequilibrium'. The question immediately arises as to how one can distinguish between a 'fundamental disequilibrium' and one that is only transsitory. To phrase the question another way, how does one recognize an equilibrium exchange rate?

In a frequently cited article Nurkse claimed that the equilibrium rate of exchange is that rate which keeps the balance of payments in equilibrium.[33] In his definition of payments equilibrium he assumes full employment and the absence of controls; he further excludes transfers of gold and other reserves plus short-term capital movements from the assessment of the state of the balance of payments. His definition of payments equilibrium is thus somewhat arbitrary. There is no reason to exclude controls (which may be imposed for a series of reasons) when defining equilibrium; equality of foreign receipts and payments is enough, provided it occurs in the absence of open unemployment. As Machlup has stated,[34] equilibrium is

[33] 'Conditions of International Monetary Equilibrium', reprinted in American Economic Association, *Readings in the Theory of International Trade*, Ch. I.

[34] 'Equilibrium and Disequilibrium: Misplaced Concreteness and Disguised Politics,' *Economic Journal*, 1958.

defined as a group of variables—in a model—so adjusted to one another as to preclude change, i.e. changeless compatibility of a set of variables. One could not possibly recognize an equilibrium in the real world; indeed, there is no such thing as an equilibrium exchange rate. *Any* rate could be an equilibrium rate depending upon the national income, rate of interest, level of employment, extent of controls, level of wages, the distribution of income, etc. Thus the Charter of the IMF does not provide a very useful policy guide as to when devaluation is appropriate.

The Elasticities Approach to Devaluation

Discussion of deficits and surpluses on current account, and the policy steps necessary to correct the imbalances, traditionally has centred on the question of demand elasticities. It long has been thought, for example, that an 'unfavourable balance of trade', i.e. a situation in which imports are greater than exports, could be remedied best by devaluation. The act of devaluation was expected to lead to a reduction in the foreign exchange price of exports and an increase in the local currency price of imports. These 'price effects' would serve to stimulate exports and reduce imports and thus bring the foreign balance back into 'equilibrium'. This procedure always would work, provided the 'stability conditions' were met.

Using the technique of partial equilibrium analysis, it was postulated that for devaluation to work it was only necessary that the sum of the domestic elasticity of demand for imports and the foreign elasticity of demand for exports be greater than one. This formulation of the stability requirements is well known as the Marshall-Lerner Theorem. It assumes implicitly that the balance of payments starts from a position of equilibrium: if initial imports were considerably greater than exports, a small decrease in the former plus a small increase in the latter resulting from a devaluation still might leave the current account in deficit; indeed, the balance could even become worse.[35] The Theorem is a correct statement of the necessary and sufficient conditions if supply elasticities are infinite.[36]

If, however, the supply elasticities are not infinite the Marshall-Lerner Theorem may not be a necesary condition for stability. Joan

[35] The elasticities analysis was extended to cover situations in which the initial trade balance was not zero by A. O. Hirschman, 'Devaluation and the Trade Balance', *Review of Economics and Statistics*, 1949.

[36] The Theorem implies that if the sum of the demand elasticities is less than one, a country can improve its trade balance not by depreciating its currency but by appreciating it.

Robinson has shown[37] that if the elasticity of demand for exports is less than one, the increase in the value of exports will be greater the smaller is the domestic elasticity of supply. 'Thus, when the foreign demand has less than unit elasticity, the maximum possible rise in the value of exports is that which is brought about when their elasticity of supply is zero.'[38] Similarly, if the demand for imports is less than unity, the value of imports will rise less, in home currency, the smaller is the foreign elasticity of supply of imports.

The Spurious Precision of a Tautology

The elasticities approach has come under increasing criticism.[39] It has become recognized that altering the exchange rate affects changes in the volume of exports and through this the amount and distribution of income, the level of employment, and the terms of trade. Thus the use of elasticities in policy formulation is rather limited. First, the calculation of elasticities at any particular moment of time may be very difficult on empirical grounds. Second, the elasticities approach does not permit consideration of the state of employment—an important factor determining the success of a devaluation. Third, the traditional analysis neglects variables other than price-quantity relationships and thus greatly over-simplifies the problem. Finally, elasticities are interdependent, and not independent of one another as the theory implicitly assumes. Suppose, for example, that a country with a low elasticity of demand for imports devalues at full employment. After devaluation *more* will be spent on imports out of any given income. Domestic factors of production therefore will be released and more will be available for exports, thereby raising the elasticity of supply of exports.

In our opinion the elaboration of the elasticities approach achieves a spurious precision. This is because the elasticities are not constant; the sensitivity of economic actors to changes in the exchange rate depends upon the circumstances of the devaluation. In Chile, for example, the foreign demand for copper depends not only upon the Chilean exchange rate and the relative price of aluminium; it also is affected by cyclical demand conditions in the importing countries, the cartel arrangements of the industry, and the strategic military situation. The foreign 'elasticity' of demand for copper will vary

[37] 'The Foreign Exchanges', American Economic Association, *op cit.*, Ch. IV.

[38] *Ibid.*, p. 89.

[39] See, for example, the two articles by T. Balogh and P.P. Streeten in the March and April 1951 issues of the *Bulletin* of the Oxford University Institute of Statistics.

as the underlying political, institutional and economic conditions change. Similarly the foreign elasticity of supply of exports will depend in part upon employment conditions in the exporting country. Looking at the import side, it is clear that the presence of quantitative controls, by excluding marginal buyers, reduces the domestic elasticity of demand for imports. Changes in the type and level of controls will have important influences on the 'elasticity'. Finally, the domestic elasticity of supply of exports will depend not only on current prices but, more importantly, on price expectations. There is no reason to believe that devaluation will encourage production for export unless present and potential producers expect that the competitive fillip will be reasonably long lasting, that is, long enough for entrepreneurs to enter the industry, obtain a profit, and amortize their investment. Thus the sensitivity of supply and demand to changed prices is not likely to be constant and therefore cannot be represented by elasticities. In considering the response of exports and imports to an altered exchange rate we must take into account such factors as the distribution of income, tax policy, monetary policy, tariff policy, quota policy, other controls, the level of employment, price and income expectations and political events. The point is that every difference makes a difference; a change in one policy parameter influences nearly all sectors of the economy. The foreign trade sector is not an appendage to the economy, but is an integral part of it; most of the sectors are interdependent and, thus, all important economic policy decisions will have to be made simultaneously. A devaluation will be a poor solution to a balance of payments problem unless other policy measures simultaneously are adopted to ensure that the desired effects in fact are achieved.

The Inappropriateness of the Traditional Categories

Sir Roy Harrod has suggested that there are four methods of correcting a trade inbalance.[40] A combination of unemployment and an external surplus should be corrected by expansionist policies. Inflation plus a deficit should be corrected by deflation. Devaluation is urged for a country suffering from unemployment and a deficit, while revaluation is recommended for countries experiencing inflation plus a trade surplus.

According to the orthodox analysis, at less than full employment devaluation will increase output and raise incomes; if the propensities to consume and invest together are less than one, some of the increase in output can be exported; the change in relative prices is

[40] *International Economics*, Ch. VIII.

expected to reduce imports. Devaluation at full employment will cause inflation and consequently is not recommended. Import restrictions are not recommended either. It is believed that restrictions will only increase 'absorption', i.e. expenditure on national consumption and investment goods, by switching money income previously spent on imports to domestic goods and services. Therefore the orthodox economist suggests, to use Harry Johnson's terminology,[41] that at full employment we need appropriate expenditure-reducing policies, and below full employment we need appropriate expenditure-switching policies, i.e. devaluation, import controls, or expansionist programmes.

There are two important difficulties with these simple policy recommendations. First, full employment is not a firm line but a rather wide band which can vary considerably. This is true in the industrialized countries and even more so in the less developed countries. Prices often begin to rise considerably before a point of full employment is reached, for the reasons analysed in Part 1 of this chapter. Secondly, many of the Spanish American countries do not fit neatly into one of Harrod's four categories. It is not unusual for one of these countries simultaneously to experience a balance of payments deficit, inflation, unemployment, inelastic supply in a few crucial sectors and slow growth. The simple Harrod-Johnson solutions are not very applicable under these conditions: deflation aggravates the problems of slow growth and unemployment, while devaluation only accentuates the inflationary spiral.

The Consequences of Devaluation

Much of the discussion of balance of payments problems in Spanish America seems to be removed from its inflationary context. It makes no sense to consider balance of payments problems in a vacuum, that is, to formulate exchange rate policies *as if* there were no inflationary problem. As long as Spanish America continues to experience secular price increases no foreign trade policy, by itself, will be completely satisfactory, although some policies will be worse than others. The trade deficit and inflation, of course, are interdependent. Nonetheless, one can say with confidence (a) the primary source of inflation lies outside the foreign trade sector; (b) sustained price increases bear a major responsibility for creating the severe balance of payments problem; (c) the resulting devaluation gives an important secondary impulse to the inflationary spiral. Thus one cannot avoid inflation by refraining from devaluing the currency, nor can

[41] *International Trade and Economic Growth*, Ch. VI.

one avoid devaluation (in one form or another) unless one simultaneously copes with inflation.

In Chile prices rose 18·4 per cent in the twelve months preceding the October 1962 devaluation. In the nine months after devaluation prices rose by 35·8 per cent. A similar process occurred in Argentina during the same period. The peso was devalued by about 50 per cent in early 1962. This led to an increase in prices of approximately 35 per cent in the first six months of the year; at the same time the supply of money increased by only 15 per cent. In Colombia between March 1951 and September 1965 there were five increasingly unsuccessful devaluations. For instance, in November 1962 the currency was devalued by 37·3 per cent and, largely as a result of this, prices rose by 31·8 per cent over the twelve months which ended in October 1963.[42]

Quite clearly, under these circumstances the Spanish American economies cannot respond well to devaluation. This does not mean that the 'stability conditions' generally are absent. No doubt 'the sum of the elasticities is greater than one', but they are not much greater. For a change in the exchange rate easily to be effective the demand elasticities must not only be 'greater than one' but they must be considerably greater.

Evidently the demand for imports in Spanish America is not very sensitive to price changes. The generally high tariffs, the import deposits, quotas, and the prohibitions on importing certain luxury items—e.g. many consumer durables and, especially, automobiles—have excluded marginal importers and greatly reduced the elasticity of demand. Devaluation and further tariff increases can only have limited effects on reducing the demand for foreign exchange. The export side provides even less grounds for optimism. Several of Spanish America's exports have been stagnating for many years. There are various reasons for this—declining world prices (e.g. coffee), increased competition from other (particularly African) producers, outmoded land tenure institutions, technological substitution by synthetics, etc.—but part of the responsibility must be borne by the persistently overvalued currencies. Exports have not expanded rapidly because it was not profitable to do so. They will not expand in the future unless conditions are such as to encourage entrepreneurs to invest in export and potential export industries. The periodic devaluations have not altered entrepreneurs' assessment

[42] See J. Sheahan, 'Imports, Investment and Growth: Colombian Experience since 1950', Center for Development Economics, Williams College, Research Memorandum No. 4, mimeo., September 1966.

of the future because it was obvious to everyone that inflation soon would eradicate any competitive advantages that might have been attained. Devaluation in the context of chronic inflation provides no incentive for investment in exports for the simple reason that profit prospects are not altered. The solution to Spanish America's export stagnation lies not in devaluation but in investment, in productivity increases. The problem is that investment in export industries will not increase unless (i) inflation is eliminated or (ii) producers are compensated for its effects either through a floating exchange rate (with all the dangers this entails) or a subsidy scheme. In the absence of policies which attack the basic export problem a devaluation only serves to increase the profits of foreign owned companies producing for export or, as in Argentina, to redistribute income to landowners. The foreign companies receive a windfall gain; there is no increase in production unless the changed cost and profit situation is deemed to be permanent. True, government tax revenues increase, but unless the marginal tax rate is 100 per cent (which it is not) there is still a net capital outflow in the form of repatriated profits.

Neither is there reason to think that devaluation significantly reduces capital flight. The capital outflow that Spanish America suffers is not due essentially to exchange rate speculation but to political uncertainty, fears that taxation will be seriously imposed and low investment yields. Inflation and the resulting destabilizing speculation may accentuate the problem but their primary influence is over the *timing* of the outflow, not necessarily over the total *volume* of the outflow. Thus we cannot expect an improvement in the balance of payments to come from the cessation of capital flight. Finally, we must not expect that devaluation automatically will reduce the level of unemployment. If exports fail to increase there will be no foreign trade multiplier effects on income, and hence, there will be no increase in employment. If the devaluation fails to encourage the formation of import substitute industries, as usually occurs, there are no forthcoming increases in domestic investment and employment. If tight credit policies are pursued unemployment may even increase. In short, what happens to unemployment largely depends upon internal policies.

Thus it appears that a policy of devaluation, unless it is reinforced by other measures, is doomed to failure. Imports usually do not fall very much; export stagnation may continue; capital outflow persists; and unemployment remains essentially unchanged. The foreign owned companies enjoy a windfall gain and the entire economy suffers from a strong impetus to inflation. The devaluation frequently is used by the government as an excuse to raise the administered prices of public utility companies. Monopolists then use this price increase as a justification for raising their prices. The increased price of imports leads the working classes to demand higher wages. This, in turn, leads to a general increase in prices, and, most importantly, to the formation of very elastic price expectations. A devaluation, far from being a stabilizing agent, seriously can disrupt an economy which is unable to withstand a sudden and strong shock.[43]

The Balance of Payments and Domestic Expenditure

Exchange rate policy in Spanish America perhaps would be less frustrating if policy makers would concentrate less on the price effects of devaluation and more on internal adjustments in expenditure. That is, more emphasis should be placed on income effects, on savings and on investment. We know that the components of total savings are domestic investment (I_d) and foreign investment. The latter corresponds to the trade surplus, the differences between exports (X) and imports (M). Thus we have $S = I_d + (X - M)$. This identity can be re-arranged so as to focus on the balance of trade: $S - I_d = X - M$. This last expression indicates that any policy which hopes to improve the balance of payments must, at the same time, increase savings relative to home investment. Thus the efficacy of a devaluation is to be assessed not in terms of elasticities but in terms of the ability of a changed exchange rate to increase savings.

If the devaluation is able to raise incomes, it will increase savings; if the devaluation is able to shift the distribution of income to savers, it will increase savings; if the devaluation is able to discourage consumption it will increase savings. If it is unable to do any of these things the devaluation will be a failure; if savings do not increase relative to domestic investment the balance of trade will not improve.

By formulating the problem in this way the intimate connection between measures to combat inflation and measures to improve the balance of payments becomes evident. It also becomes evident why

[43] Everything else being equal, the larger the proportion of imports to national income the lower the likelihood that devaluation will correct a balance of payments deficit. The reason for this is that the greater is a country's dependence on foreign trade—i.e. the larger is the proportion of imports to domestic costs and domestic expenditure—the greater will be the impact of a given devaluation on raising domestic prices and accelerating inflation.

devaluations generally have not been successful. First, incomes do not rise because exports fail to expand, for reasons already discussed. Second, inflation temporarily may make the distribution of income a little more regressive, while the groups which are harmed build their defences, but it is unlikely that savings will noticeably increase since the wealthy groups have a high marginal propensity to consume. Finally, any influence the 'real balance effect' or money illusion may have on reducing consumption is rapidly eradicated by the formation of elastic price expectations. Ultimately, it is expectations which destroy the ability of a change in the exchange rate to increase savings: poor profit expectations discourage the expansion of export industries; expectations of continuing inflation prompt the construction of defences to avoid changes in the distribution of income; elastic price expectations encourage consumption at the expense of savings. Only by attacking expectations and the 'real' phenomena which underlie them can savings be increased. Yet an increase in savings is essential for eliminating inflation and the balance of payments problem.

In summary, the effects of a devaluation are very difficult to assess. Even when a country is experiencing unemployment and a deficit a devaluation may not be successful if inflation soon wipes out the competitive gains on the export side. In fact, exchange rate depreciation will be unsuccessful even if entrepreneurs only *believe* that inflation will quickly eliminate the competitive advantages thereby attained. Thus it is very important to consider expectations and the potential rate of growth of output before advocating devaluation. Furthermore, short and long-run effects must be distinguished. A devaluation may be reasonably successful in the short-run, or better still in the medium-run, but by encouraging inflation it may be self-defeating in the long-run.

The identity expressed by the last equation can be as misleading as the elasticity formulae. This is because the *effects of policy will not be symmetrical* in the short- and long-runs. The balance of payments can be improved in the short-run either by increasing savings or by reducing domestic investment, i.e. by deflation. In the long-run, however, the state of the balance of payments will depend in large part on the level of investment and the relative rate of increase of productivity. That is, *given* the level of effective demand, the balance of payments will be worse the lower the rate of capital accumulation. Thus, in the long-run, it is essential to increase investment, and not reduce it. The short-term difficulties can be met, not by deflation or devaluation, but by imposing temporary controls until more permanent solutions have time to take effect.[44] That is, one must not consider balance of payments problems in isolation but must continually refer to the general economic conditions of the society. Frequently a series of measures may be required to achieve the given (and multiple) ends.

A Policy Mix

The question remains: what in the short-run, if anything, can be done primarily in the foreign trade sector to improve the foreign exchange position of the individual Spanish American countries? A brief answer would be 'very little'. There is, however, an approach which as yet has not been fully exploited. This consists of a more intelligent use of quota, tax and subsidy schemes or, alternatively, of multiple exchange rates.

In the Spanish American republics, where the demand for imports is highly inelastic and the supply is highly elastic, it is possible that better results could be obtained by a quota system than by tariffs. A quota fixes the quantity of imports and lets the internal price adjust; a tariff, assuming the elasticity of supply is infinite, sets the price of imports and lets the volume adjust. Since we are really interested in controlling the *value* of imports (price times quantity) one would think that a tariff or quota scheme would be equally satisfactory; the choice between the two could be based on administrative convenience or political prejudice. If, however, the demand is inelastic, controlling the quantity of imports will largely control their value. That is, the supply elasticity will determine the external price; the quota will determine the quantity; and the value will thereby be controlled. A tariff policy, on the other hand, will only influence the internal price; the volume, and hence the total value of imports, will be undetermined.

Thus a quota system has certain advantages over tariffs. It would

[44] Before going on to defend in more detail the need for trade controls, we should mention briefly that more and more frustrated economists in Spanish America are beginning to advocate 'floating' exchange rates. In conditions of secular inflation such a position represents advocacy of continuous and permanent devaluation. Therefore all the arguments against devaluation apply here. In addition, however, there are other disadvantages peculiar to this policy: (1) the danger of continuous destabilizing speculation; (2) the danger of sharp fluctuations in the terms of trade (since short-run 'elasticities' are less than long-run 'elasticities'); (3) the increased risks of trading and lending; (4) the interference of fluctuating rates with direct controls (and vice versa); and (5) the great danger that continuous depreciation will accelerate inflation through the effect of rising import prices on wages and hence on all other prices.

be unwise, however, for the government simply to give the import permits away. An intelligent government would not do this, but instead would auction them off to the highest bidder. This technique would eliminate the windfall gains of a few lucky importers and greatly increase government revenues.

The quota system would only partially alleviate the balance of payments problem. If the auction revenues merely flowed into the Treasury coffers government savings would greatly increase, but the export performance would be severely damaged. A better policy would be to use these revenues to subsidize selected export industries whose expansion is deemed beneficial to the economy as a whole and whose expansion is being slowed down by the over-valued exchange rate. In effect, a quota and subsidy scheme of this sort would constitute a devaluation of the currency—with two important differences. First, the government is assured that the foreign exchange value of imports will fall. Secondly, windfall gains to foreign-owned export companies can be avoided. The export subsidies would be discriminatory measures similar to 'dual' or 'multiple' exchange rates; the 'devaluation' would be selective and would assist only those firms or industries which need it in order to expand production and sales.

In countries where the productivity of labour and the returns to investment differ widely from one sector to another the arguments in favour of a quota, tax and subsidy scheme, as outlined above, or a multiple exchange rate system, are powerful.[45] A uniform exchange rate system cannot be appropriate, unless corrected by special taxes and subsidies, if the economic characteristics of the various sectors of the economy are not uniform.

Although Colombia has not adopted an exchange rate system identical to that being advocated, her experience illustrates the way a discriminatory control system can be used.

Colombia's most important export, coffee, provides about two-thirds of the country's foreign exchange earnings. Unfortunately, coffee exports cannot be expected to increase very rapidly. Indeed, the value of coffee exports in 1967 were somewhat lower than they were a decade earlier, despite the existence of the international coffee agreement. The second largest export, petroleum, similarly cannot be expected to increase very rapidly. Thus the two products which at present account for nearly 80 per cent of total exports cannot be relied upon to provide adequate earnings of foreign exchange in the

[45] See N. Kaldor, 'Dual Exchange Rates and Economic Development', *Economic Bulletin for Latin America*, September 1964, where a special exchange rate to encourage the exports of manufactured goods is advocated.

future. In other words, most of the additional exports which Colombia needs must come from products other than coffee and petroleum, i.e. they must come from the so-called minor exports.

The authorities are aware of this situation and exchange rate policies designed to stimulate exports are being introduced. In March 1967 two exchange markets were established: an exchange certificate market and a capital market.

The capital market rate is fixed at 16·25—16·30 pesos per US dollar and applies to trade in invisibles and capital transactions. The exchange certificate rate, however, is a fluctuating one. It applies to major imports. Exports other than coffee, petroleum and cattle hides also are traded at the certificate rate, and in addition, receive a 15 per cent subsidy.

This is clearly an attempt to increase the volume of minor exports. Minor exports have, in any case, been rising rapidly in the last few years, *viz.* from $70 million in 1960 to $128 million in 1966, or from 11·3 per cent of total exports to 23·2 per cent. Moreover, studies have shown that the minor exports are quite responsive to variation in the exchange rate, *provided devaluation does not result in rapid inflation.*[46] Thus, one can anticipate that the combination of a flexible exchange rate and a 15 per cent subsidy should lead to a further rise in these exports. Some observers expect minor exports to equal $300 million by 1970. This is probably too hopeful, but $200 million is certainly possible.

The former coffee exchange rate was abolished and the exchange differential was replaced by a variable (and declining) export tax. So coffee today is traded at the certificate market rate after adjusting for the export tax, which ultimately will decline from 19 per cent in December 1967 to 16 per cent in December 1968. This provides a double stimulus to coffee exports. First, coffee is given a more favourable exchange rate and this exchange rate is itself depreciating over time. Secondly, the export tax which replaces the former exchange differential is scheduled to decline gradually each month until December 1968. Given that Colombia already is producing more coffee than can be sold and that stocks are accumulating at a rate of 700,000 bags a year, the new coffee policy appears to be very unwise. The only satisfactory explanation for the new policy is that the authorities, under pressure from the International Monetary Fund, felt obliged to move as far as possible toward a single exchange

[46] J. Sheahan and S. Clark, 'The Response of Colombian Exports to Variations in Effective Exchange Rates', Research Memorandum No. 11, William College, June 1967, mimeo.

rate system, and to gradually abandon a system whereby taxes on coffee exports could be used to subsidize non-coffee exports.

Finally, there is a fixed exchange rate of 7·67 pesos to the dollar for some petroleum transactions.

In summary, there are six exchange rates now operative in Colombia:

(1) a fluctuating major imports rate = the certificate rate, quoting at about 15·28 to the dollar in September 1967.
(2) The minor exports rate = the certificate rate plus a 15 per cent subsidy.
(3) The new coffee rate = the certificate rate adjusted for a declining export tax.
(4) The capital market rate of 16·25–16·30 to the dollar.
(5) The petroleum rate of 7·67 to the dollar; and
(6) A black market rate of about 17·00.

Obviously an exchange control system such as this is quite complex. Yet the Colombians are able to administer it competently. Moreover, the control system has enabled the authorities to apply discriminatory policies—stimulating production where prospects were bright and discouraging exports when supply was in danger of becoming excessive. In this way, foreign exchange earnings could be substantially increased. It is a great pity that the government felt compelled to modify this system and, by narrowing the differential between the coffee rate and the certificate rate, move towards a more unified and orthodox exchange rate system.

APPENDIX

Given that wage rates are annually readjusted to compensate for the price increases which occurred over the previous year, it is easy to demonstrate that a reduction in the rate of inflation achieved through a contraction of the money supply will increase the government's deficit or reduce public savings.

Let α = the proportion of money income (Y) paid in taxes and π = the rate of increase of prices. We assume that $\pi_{t-1} > \pi_t$, i.e. that the rate of inflation has declined between the two years.

The government deficit is equal to expenditures minus receipts, or

(1) $D_t = E_t - R_t$.

Receipts depend upon the current level of money income, i.e.

(2) $R_t = \alpha Y_t$.

Let us now assume that the level of income is constant in real terms, i.e.

(3) $Y_t = Y_{t-1} + \pi_t . Y_{t-1} = (1 + \pi_t) Y_{t-1}$

Government expenditure consists of payment of salaries and purchases of materials.

(4) $E_t = S_t + M_t$

Assume that the purchase of the physical volume of materials is constant so that

(5) $M_t = M_{t-1} . \pi_t + M_{t-1} = M_{t-1} (1 + \pi_t)$

We also assume that the number of government employees (N) is constant but that the wage rate (W) is escalated or readjusted to compensate for the previous inflation.

(6) $S_t = N_{t-1} [W_{t-1} + W_{t-1} . \pi_{t-1}]$
$= N_{t-1} [W_{t-1} (1 + \pi_{t-1})]$

Substituting equations (2) through (6) into (1), we can obtain a new expression for the government's current deficit as follows:

(7) $D_t = N_{t-1} [W_{t-1} (1 + \pi_{t-1})] +$
$M_{t-1} (1 + \pi_t) - \alpha[Y_{t-1} (1 + \pi_t)].$

Since all physical quantities are constant from year $t-1$ to year t, the only difference being that the rate of inflation is greater in the former than in the latter, the final equation demonstrates that stabilization will acentuate the deficit by the amount

$$N_{t-1} [W_{t-1} (\pi_{t-1} - \pi_t)].$$

This amount can be called the wage excess (X_w). Assuming the volume of government purchases—including purchases of labour—is constant, the increase in the deficit can be eliminated only by an autonomous increase of real income, e.g. by a rise in exports, of X_w/α—or by raising α by the amount X_w/Y_t.

CHAPTER VI

Regional Integration

The Spanish speaking countries of South America have been only partially successful in processing their traditional primary exports or in developing entirely new industrial exports. At the same time, further expansion of exports of raw materials and foodstuffs is limited by the low price and income elasticities of demand. It would appear, then, that additional efforts will be required to expand exports of manufactured goods; it is only in such a manner that Spanish America can solve its long-run export problem and obtain the capital goods necessary to sustain a high rate of investment.

One of the explanations why underdeveloped countries such as those in Spanish America do not up-grade their primary products into more elaborate exports is that tariff rates in industrialized countries are positively correlated with the degree of product elaboration. We know, for example, that the US tariff on frozen fish is much less than the tariff on pre-cooked fish. This hampers the development of the Peruvian, Chilean and Ecuadorian fishing industries. Similarly, the tariff rates become increasingly higher as one moves from iron ore to raw steel to industrial machinery. This delays the creation of an iron and steel complex in Colombia, Argentina and Chile. Thus tariff discrimination against products with high value added content tends to force the underdeveloped countries to remain primary producers by placing them increasingly at a competitive disadvantage with domestic producers in the advanced economies. A simple model will demonstrate this.

Let c_1, c_2, c_3 = the cost of stages one, two, three respectively;
X_1, X_2, X_3 = total cost of production elaborated through stages one, two, three respectively;
t_1, t_2, t_3 = tariff rates applied at the first three stages. We assume $t_1 < t_2 < t_3$.
P_1, P_2, P_3 = selling price abroad, including tariffs, of the commodities.

Thus $X_1 = c_1$ and $X_2 = c_1 + c_2$, etc.

Similarly, $P_1 = X_1 + X_1t_1 = c_1 + c_1t_1$ and

$$P_2 = X_2 + X_2t_2 = (c_1 + c_2) + (c_1 + c_2)t_2, \text{ etc.}$$

The additional selling cost of up-grading the raw material one stage ($P_2 - P_1$) must be less than the cost of production in the importing country. For if not, the process of elaboration would not be profitable. If we let W_2 represent the cost of producing stage two in the advanced economy, the condition of profitability can be represented as

$$\begin{aligned} P_2 - P_1 &= [X_2 + X_2t_2] - [X_1 + X_1t_1] \\ &= c_2(1 + t_2) + c_1(t_2 - t_1) \leqq W_2 \end{aligned}$$

Thus it becomes progressively more difficult for less developed economies to compete over the tariff barrier because the higher tariffs apply not only to the last stage but to each of the earlier stages of manufacturing as well.

The table below presents US full tariff rates for several primary commodities and their subsequent stages of elaboration.[1] All of these primary products are important exports of one or more Spanish American countries.

Bela Balassa has studied the structure of tariffs in the capitalist industrial countries. In general, excluding United Kingdom imports from the Commonwealth, he found that in 1964 the effective rate of protection on 'simply processed' goods was 22·6 per cent; the effective rate of protection rose to 29·7 per cent on 'further processed' goods and to 38·4 per cent on 'fully manufactured' goods.[2] At the same time it was found that 71·2 per cent of the imports from the underdeveloped countries were in a 'crude form', while only 2·1 per cent were 'fully manufactured'. This pattern of trade, quite clearly has been greatly influenced by the system of tariffs adopted in the industrial economies.

This commercial discrimination by the wealthy countries against the industrialization of the poorer nations appears to be reinforced by the policy of common carriers to charge discriminatory freight rates. As A.N. McLeod points out,[3] 'typically, bulky raw materials

[1] For additional information see UN, ECLA, *Latin America and the United Nations Conference on Trade and Development* (E/CN/12/693), February 20, 1964, p. 112; Political and Economic Planning, *Atlantic Tariff and Trade*, 1962; Bela Balassa, *The Structure of protection in the industrial countries and its effects on the Exports of Processed Goods from Developing Nations*, UNCTAD document (TD/B/c. 2/36).

[2] B. Balassa, *ibid*.

[3] 'Trade and investment in Underdeveloped Areas: A Comment', *American Economic Review*, June 1951, p. 416.

Table VI: 1

U.S. TARIFF RATES ON SELECTED COMMODITIES

Product	US tariff rate
(1) Fish products	
(a) fish meal	free list
(b) fresh or frozen fish	1¢–3¢ per pound
(c) pre-cooked fish	30–40% ad valorem
(2) Wood products	
(a) wood pulp	free list
(b) paper press	30% ad valorem
(c) rayon	45% ad valorem
(3) Leather products	
(a) hides	10% ad valorem
(b) leather	12·5–30% ad valorem
(c) shoes	20% ad valorem
(4) Wool products	
(a) raw wool	37¢/lb. + 20% ad valorem
(b) wool yarn	40¢/lb. + 35,45 or 50% ad valorem
(c) manufactures	50% ad valorem
(5) Iron and steel products	
(a) iron ore	free list
(b) pig iron	$1·12 per ton
(c) bar iron	$20–$30 per ton
(d) industrial machinery	
(i) mining machines	27·5% ad valorem
(ii) machine tools	30% ad valorem
(iii) textile machinery	40% ad valorem
(6) Grape products	
(a) grapes	0·5–1¢/lb
(b) non-alcoholic grape juice	70¢/gallon
(c) wine	70¢/gallon + $5 per gallon of alcohol
(d) brandy	$5/gallon

are accorded favourably low rates, whereas manufactured and highly processed goods pay much higher rates'. Thus government tariff policy combined with monopolistic pricing by international shippers introduces a very strong bias against the industrialization of the underdeveloped economies. Plants and factories are located not near the source of raw materials, i.e. in the underdeveloped countries, but near the final markets, i.e. in the advanced economies.

Even in the absence of tariff and freight rate discrimination it is unlikely that Spanish America could become industrialized by exporting manufactured goods to the advanced economies. At the moment, these nations are unable to compete with Europe and North America in the production of most industrial products. Their static comparative advantage, as measured by market prices, does not lie in these goods. More important, without substantial preferential treatment, these nations are not likely to *acquire* a comparative advantage in manufactured goods in the near future. The Spanish American countries have a demonstrated incapacity for transforming their economies, by altering the composition of output and demand, in the face of changes in relative prices. This transformation problem is due essentially to the lack of a vigorous industrial entrepreneurial class. This does not imply that there are no risk takers; it means only that those who *are* willing to assume risks do not enter industry. It is extraordinarily difficult for an entrepreneur from an underdeveloped country to compete with an investor from an industrial nation. This is true whether the competition takes place in the underdeveloped economy (as argued in Chapter III) or in the domestic market of the industrial economy. Thus for all these reasons, Spanish America was forced to select a development strategy which did not rely on exporting manufactured goods to the United States and Europe.

The elected strategy, strongly recommended at the time by the United Nations Economic Commission for Latin America,[4] was one of import substitution at the national level. In the early stages and up until the mid-1950s the policy encountered some success—particularly in Argentina, Chile, Mexico, Brazil, and to a certain extent even in Colombia. Industrial plants, designed to supply the domestic market of each of the Republics, were established behind very high tariff walls. In recent years this policy has been carried to excess and has had unfortunate repercussions. The internal markets in these countries frequently are too small even to justify the erection of one efficient sized plant. Monopolies with strong vested interests, uncontrolled by the state, were created. Prices of manufactured products rose substantially and affected a redistribution of income toward the monopolistic producers. There was little chance that these 'infants' would ever grow up, and in any case, the limited market soon stopped this 'development from within'.

[4] See R. Prebisch, 'Commercial Policy in Underdeveloped Countries', *American Economic Review*, May 1959; UN, ECLA, *El desarrollo Economico de Latino-America y sus Principales Problemas*, 1950.

Moreover, there are inherent difficulties in a strategy of import substitution for small nations. Such a strategy is almost inevitably biased, first, in favour of small plants and, second, toward the consumer goods industries. The first bias tends to lower the level of income by creating inefficiency[5] and the second bias tends to retard the rate of growth. Indiscriminate protection of consumer goods industries usually is combined with liberal import policies for raw materials and capital goods. This tends to discourage export promotion as well as backward linkage import substitution into capital goods, yet one or both of these activities is essential for sustained rapid growth.

Apart from adding to the stocks of consumer goods, an economy, physically, cannot undertake investment without machines. The necessary machines can either be imported from abroad or produced locally. The former requires a rapidly expanding export sector and the latter a domestic capital goods industry; if neither exists, a high level of investment may be impossible. Furthermore, a high level of investment must be *financed* by restraining consumption. In many Spanish American countries it seems that the policy of import substitution, by emphasizing the production of consumer goods in relation to domestic demand and output, has undermined the restraints on the consumption of consumer goods. In other words, as import substitution of consumer goods proceeds there is an automatic de-control of consumption due (a) to the increased availability on the market of scarce consumer goods and (b) to the fall in government revenue and public savings associated with the shift in supply of consumer goods from heavily taxed imports to lightly taxed domestic production.[6]

Despite a policy of import substitution, balance of payments problems may persist. This may compel the government to re-impose controls over imports. If imports of raw materials and spare parts are reduced excess capacity in the domestic consumer goods industries will be created. This may tempt the authorities to try to obtain rapid growth by utilizing excess industrial capacity. They may then reduce imports of capital goods in order to increase imports of inputs used by the consumer goods industries. This, of course, will generate a temporary boom, but it will do so at the cost of a lower

[5] In some instances highly protected manufacturing activities have a negative value added. For the case of Pakistan see R. Soligo and J. Stern, 'Tariff Protection, Import Substitution and Investment Efficiency', *Pakistan Development Review*, Summer 1965.

[6] See J. Power, 'Import Substitution as an Industrialization Strategy', mimeo., Williams College, Williamstown Mass., January 1966.

rate of capital accumulation and a slower long-run rate of growth. Thus the government will be in a dilemma: either it must accept considerable excess capacity or a low level of investment. It is at this point that policies of import substitution of consumer goods cease to be viable.

Recognition of these difficulties has forced the Spanish American governments to reconsider their development strategy. The new tactic is to establish a regional free trade area. Before discussing this new policy as it applies in Spanish America it appears worth-while to consider generally the economic theory of customs unions in a development context. Accordingly, the remaining portion of this chapter is divided into five parts. Part 1 consists of a short section describing optimum and sub-optimum conditions and the relationship of the theory of the 'second-best' to the theory of customs unions. In Part 2 we consider the theoretical sources of an increase in economic welfare resulting from the creation of various types of integration. Both once-for-all and continuous gains are treated and their relevance to Spanish America is discussed briefly. Part 3 is devoted to a discussion of the problem of dynamic adjustment and the policy steps necessary to prevent each of the member nations from experiencing a decline in welfare. Thus the focus is on the distribution of the benefits of integration. In Part 4 we consider some of the important details and weaknesses of the treaty which established the Free Trade Area, and in Part 5 we indicate some of the steps that are necessary to accelerate regional integration.

1 CUSTOMS UNIONS AND THE 'SECOND-BEST'

Given two assumptions, *viz.* (i) the international immobility of factors of production and (ii) the condition in all countries that prices of goods equal their marginal social costs and this in turn equals the consumer's marginal social valuation of goods, it can be claimed accurately that free trade will provide the marginal conditions necessary for the achievement of economic efficiency. Under these circumstances trade will be optimized when the ratios of prices on all commodities are the same in all countries. Maximum production simultaneously will be achieved when the ratios of marginal social costs of all commodities are the same in every country. The satisfaction of the above conditions is enough to ensure that production and effort will be at the optimum. Production will take place on, and not within, the transformation curve. Thus free trade is the allocation ideal. It is *the* optimum solution to international efficiency.[7]

In the real world, of course, there are commonly considerable divergences between marginal values and costs. Paradoxically, it is often true that a single movement toward free trade, i.e. a reduction in the divergence between marginal value and cost, will not lead to an increase in economic efficiency and welfare, but may reduce it. This is so because we usually change one policy at a time. Previous changes become, for the purpose of this decision, parameters. It then becomes possible for a reduction in a particular divergence from the optimum to lead to an increase in the rates of divergence. Given several divergences between marginal values and costs, the maintenance of one particular divergence may help to offset the negative effects of another. Perhaps the removal of both would improve welfare, but if one is going to be kept, it may be better that both should be kept. In other words, if *all* the optimum conditions cannot be achieved simultaneously, the fulfilment of *some* of the optimum conditions will not necessarily make a country better off than would their non-fulfilment.[8] The sub-optimum conditions, i.e. the maximum attainable given certain divergences, usually will require a violation of the optimum conditions. There is no *a priori* presumption that a movement *toward* the *best* position represents a movement *to* a *better* position. It all depends upon the circumstances of the particular case.

Economic integration represents an attempt of the integrating nations to maximize (or at least increase) their welfare—given the existence of distortions in their trade and the world's structure of production, the elimination of which is beyond the control of the member countries. Thus, though free trade might constitute an hypothetical optimum, if would be unrealistic to pretend that the necessary conditions for achieving the optimum are likely to be realized. Consequently, given the impossibility of reaching the 'first best' we must attempt a 'second best' solution: economic integration among a group of countries is one of the possible alternatives.

2 THEORETICAL CONSIDERATIONS

There are several degrees of economic integration and the possibilities for improved resource allocation and faster growth will

[7] It does not follow, however, that free trade maximizes the welfare either of the world as a whole or any country in particular. This will depend upon the distribution of income between (a) the various countries, (b) regions within countries, and (c) individuals within regions.

[8]. See R. G. Lipsey and K. Lancaster, 'The General Theory of the Second Best', *Review of Economic Studies*, No. 63, 1956–57.

depend in large part upon just what type of integration is anticipated. The most simple arrangement is a Free Trade Area. Members of the Area agree to the mutual reduction or elimination of tariffs, but each member country maintains separate tariffs for third countries. One of the major difficulties with this arrangement is that the relatively low-tariff member countries always are tempted to import goods from non-members for re-export to their partners. This problem usually is attacked by specifying that re-exports must have a certain percentage of value added by the member country before the good is eligible to receive favourable treatment. This device does not solve the problem completely, however, as large integrated companies with widespread international operations usually can make costs appear where they like. A better solution is the formation of a Customs Union. A customs union differs from a free trade area only to the extent that member countries agree to a common tariff level for third countries. Closer integration is achieved under a Common Market. Under this institution controls on the movement of capital and labour are eliminated so that factors as well as goods are free to move. A complete Economic Union is achieved when a common market is supplemented by regional planning and co-ordination of government policies for growth and stability. Thus the costs and benefits of integration are likely to be varied and come from many sources, depending upon the types of institutions established. The most important effects are analysed below.

The Static Production Effects of a Customs Union

It is obvious, however, that *from the point of view of static resource allocation* all of these forms of integration must be analysed in terms of the theory of the second best. The purpose of integration is to make economic conditions better, rather than to attain the best possible. The economic purpose of a customs union is not to increase trade for its own sake, but to increase the welfare of its members by reducing or removing trade restrictions.

The methods used for increasing welfare are to shift sources of supply and the composition of demand. As Viner has indicated,[9] if the formation of a union is to be a movement toward a better allocation of resources it must be predominantly a movement in the direction of goods being supplied from lower cost sources than before.

Whether the reduction of trade barriers between countries of the union will improve the efficient use of resources depends upon whether the 'trade creation effects' exceed the 'trade diversion effects'.

[9] *The Customs Union Issue.*

Trade creation involves a switch from high priced domestic production, which was protected by a duty, to a lower priced union member's production. *Assuming the economy enjoys full employment, costs of adjustment are zero, and there are no external economies,* trade creation always has positive economic benefits. Trade diversion implies a transfer of the source of trade from low cost areas outside the union to high cost producers inside the union; it tends to be uneconomic and wasteful.[10] Presumably, any customs union will have some trade diversion and some trade creation effects. The question is which of the two effects predominates.

Suppose we have eight countries—A, B, C, . . . H— and consider only one commodity, X.[11] Assume country A has the lowest cost in the production of X and the costs in the other countries rise in order until we reach country H which has the highest costs. This can be represented diagramatically.

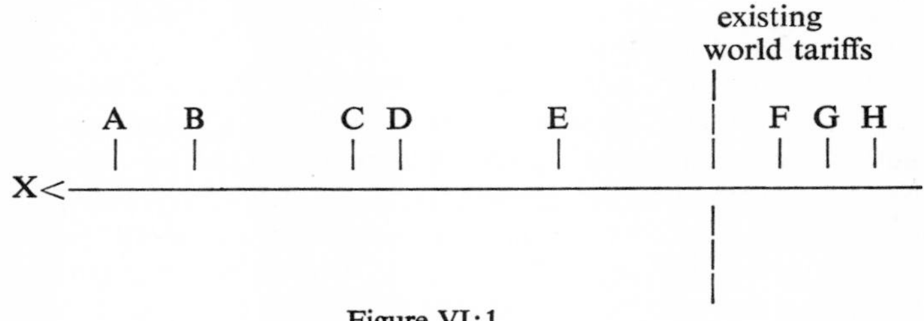

Figure VI:1

Countries F, G, and H have costs so high they cannot compete in product X even behind the existing world tariffs. These countries therefore are importing X from A. The X industry is protected in countries B, C, D, and E, and production is undertaken for domestic consumption.

A customs union may have three important effects on the production[12] of good X: (1) a union between (B, C, D, E) and (F, G, H)

[10] The neo-classical analysis, part of which we are presenting in this section, assumes full employment. Such an assumption has no applicability in Spanish America. If there is substantial unemployment there will be no real social cost of establishing industries through trade diversion, and in this case the relevance of the Vinerian argument vanishes.

[11] See H. Makower and G. Morton, 'A Contribution Towards the Theory of Customs Unions', *Economic Journal*, March 1953.

[12] It is possible for a trade diverting customs union to result in an increase in welfare by causing shifts in *consumption* expenditure from a less to a more preferred pattern. See R. G. Lipsey, 'The Theory of Customs Unions: Trade Diversion and Welfare', *Economica*, February 1957.

diverts trade from country A; (2) a union between A and (F, G, H) has no effect; (3) a union between A and (B, C, D, E) will have a positive trade creating effect; in fact, any union between countries with costs less than those of F will increase efficiency.

It is important to note, first, that in only one out of the three major cases did the formation of a customs union increase efficiency. Secondly, the increase in output and efficiency associated with the trade creation effect is due not to an increase in productivity in any particular industry but to the increased specialization in those goods in which a country has a comparative advantage. That is, the gains from trade creation can be realized only if there is a high degree of substitutability of factors and of products. Countries which have transformation problems will be unable to reap the potential allocation gains from trade creation. Specialization will be increased not by expanding one industry while contracting another, but simply by destroying the now unprotected industries and creating unemployment.[13]

Thirdly, it is important to recognize that trade diversion and creation also are relevant to tariff reductions that include some commodities but exclude others.[14] This case is particularly relevant for Spanish America because the Treaty of Montevideo, which established the Latin American Free Trade Area (LAFTA), seems to envisage a commodity by commodity approach to tariff reductions.

Assume that Chile uses fertilizers to produce wheat and first imposes a tariff of 50 per cent on both commodities, but later forms a union with Argentina and eliminates the tariff on fertilizers. The cost situation is depicted below.

Costs		Prices with 50% tariff	Prices with tariff on fertilizers removed
CHILE			
Fertilizer	50	50	50
Wheat	100	100	80
ARGENTINA			
Fertilizer	40	60	40
Wheat	60	90	90

Argenina can produce fertilizer 20 per cent cheaper than can Chile, but she can produce wheat 40 per cent cheaper. So by the principle of comparative advantage Chile should specialize in fertilizer and Argentina in wheat. This is done when a 50 per cent tariff is applied

[13] See Chapter II, pp. 93–7.

[14] See P. P. Streeten, *Economic Integràtion*, pp. 35–6 from which the example used below is taken.

to both products. After the formation of the union, however, Chile removes her tariff on fertilizers. The cost of this commodity falls by 20 per cent and, if we assume that wheat prices are proportional to fertilizer costs, food prices also fall by 20 per cent, to 80. It now becomes 'profitable' for Argentina to specialize in fertilizer and Chile to specialize in wheat. Thus the effect of the union is to divert trade to an inefficient production pattern.

Something similar to this theoretical example occurred in practice after the first annual negotiations required by the Treaty of Montevideo to establish 'National Lists'. Previously, Chile used to export semi-elaborated copper to Argentina. In 1962 Argentina removed the 9 per cent tariff on imports of copper rods but maintained the tariff on imports of semi-elaborated copper. As a result, Argentine imports of semi-elaborated copper were substantially reduced, and were substituted by imports of copper rods from Chile.

A False Generalization: the Happy Union of Complements

The neo-classical analysis as presented so far does not enable us to make any *a priori* judgments about the value of customs unions. We can assume, of course, that if all countries had prohibitive tariffs, then *any* customs union would be beneficial since 'some trade is better than no trade'. Similarly, the higher were the initial rates of duty in the uniting countries, the more likely is a customs union to increase efficiency. According to this argument, in other words, the extent of the gains from union depends upon the extent of the real cost differences between the uniting countries. The greater the cost differences, the greater the benefits from union.

This argument has been generalized to favour integration of potentially complementary economies and to oppose integration of competitive economies. That is, it is stated that the allocation gains from union will be greater for groups composed of complementary economies, i.e. countries which have potentially dissimilar production costs. The potential gains are greater if two countries which are now competing behind barriers but which have potentially dissimilar costs unite than if rival (competitive) economies, i.e. countries with similar transformation ratios, unite. If both countries have high initial tariffs the production pattern in the two will be similar. If we assume, however, that the real costs or transformation ratios differ sharply, then a tariff reduction will create trade and the final economies will be complementary.

This generalization could be used to condemn Latin American integration. The member countries in many instances are rival economies, not complementary. Their resource endowments are similar and all are labour abundant and capital scarce countries. Hence one could argue that the formation of the Latin American Free Trade Area is not likely to increase efficiency. This generalization, however, is not relevant if potential economies of scale and increasing returns exist between rival economies, for the increase in efficiency is due not to trade creation but to the expansion of the market and the exploitation of decreasing costs. As we shall see, this is probably the most important argument in favour of integration in Spanish America.

The Quantitative Relevance of the Neo-classical Theory

Relatively little effort has been devoted by economists to testing the empirical importance of the above theory. What few studies have been made, however, indicate that the net gains from trade diversion-creation are low. Professor Harry Johnson's study, for example, 'demonstrated' that the gains to Britain of freer trade with Europe would be of the order of 1 per cent of gross national product.[15] That the static allocation gains almost inevitably are low can be demonstrated with a very simple model.

Suppose that our Spanish American country has a ratio of imports to total income of 1:5. That is, 20 per cent of total income is imported. Then let us assume this country forms a customs union with several of its neighbours so that the ratio of imports rises to 3:10; i.e. 30 per cent of income is now imported. Let us further assume that this increase in trade (10 per cent more of national income is traded) represents a shift in sources of supply from dear domestic producers to low cost partner producers. Thus we assume there are no trade diversion effects. Suppose the initial tariff rate was 50 per cent *ad valorem* and the additional imports are 49+ per cent cheaper than the equivalent domestic products. Internal prices initially were 150 (100 representing the production cost abroad and 50 the tariff). The elimination of the duty reduces prices to 100, that is, they fall 33 per cent. The gain from union comes from the fact that 10 per cent of the national income now can be purchased at prices 33 per cent lower. Hence the approximate gain to the economy is 3.3 per cent

[15] See 'The Gains from Freer Trade with Europe: An Estimate', *The Manchester School*, September 1958; B. Balassa, 'Trade Creation and Trade Diversion in the European Common Market', *Economic Journal*, March 1967. See also E. A. Farag, 'The Latin American Free Trade Area', *Inter-American Economic Affairs*, Vol. 17, No. 1, Summer 1963 for a similar study of LAFTA and the false conclusion based upon it.

(1/10 × 1/3) of national income. This is a once-for-all (static) gain and represents about one year of slow growth. Hence even under our heroic assumptions the allocation gains are very low.

Shifts in the Terms of Trade

Under special circumstances a union may help to shift the terms of trade in a favourable direction. If the quantity of imports by the member countries constitutes a sufficiently large fraction of the world production of these goods, a reduction in the demand for imported goods, arising from the creation of a customs union, will cause a fall in their price. Furthermore, goods formerly exported will be absorbed by the member countries. If the supply of these goods available on the world market is substantially reduced the price of exports will rise. For such effects to occur the member countries must be major world suppliers of the exported goods and must constitute an important market for the imported goods. It is doubtful whether LAFTA meets these conditions. Hence one cannot expect the Free Trade Area to improve the region's terms of trade.

Of much greater importance are likely changes in the intra-area terms of trade. It is quite possible that the weaker members could become importers of high cost manufactured goods from the Area while remaining exporters of agricultural commodities and raw materials. This shift in the sources of supply would result in a deterioration of the terms of trade of the economies producing primary products. Bolivia at first refused to join LAFTA precisely because (i) the bulk of her primary product exports go to non-member countries, (ii) she can import manufactured goods from the industrialized countries at prices lower than can be offered in Latin America, and (iii) the Treaty of Montevideo has no mechanism for assisting industrialization in the member countries. Thus Bolivia feared, probably correctly, that the likely decline in her terms of trade would not be compensated by the possibility of establishing her own manufacturing enterprises. Thus to this extent the trade diversion argument is relevant. One solution to this problem would be to allow the weaker partners to maintain some tariff discrimination against imports from other member countries as well as, perhaps, allowing some subsidies for exports.

Capital Inflows

High tariff rates to the outside world combined with few or no tariffs between the uniting nations may provide a strong attraction to foreign capital. Outside firms may flock to the union and establish branch factories in order to get inside the protective barriers. Credit may be mobilized on a large scale. New products may be created. Furthermore, a union will encourage businessmen to increase their contacts and will facilitate the dissemination of knowledge about alternative production techniques and business methods. In addition, the establishment of a customs union may increase the bargaining power of the member nations. Subsequently it may be easier to negotiate loans, trade concessions and reciprocal tariff reductions from the advanced industrial nations.

One should not be too confident, however, that the establishment of a customs union, by itself, will lead to a large and continuing inflow of capital. Foreign capital, with a few important exceptions, seems to be attracted in large volume to *growing* markets, not just to large ones. If the union stimulates growth, then one can reasonably expect that this growth will attract capital, but if the union perpetuates stagnation, foreign capital may stay away. Even if the union is successful in attracting a large capital inflow there can be no *a priori* guarantee that this capital will flow to the regions or industries where it is most needed. Thus the member countries cannot rely on foreign capital either to initiate growth or to satisfy high priority needs.

The growth effort is an internal responsibility, and foreign funds should only be marginal, but potentially complementary, to the domestic and regional savings efforts. This does not mean that capital imports should be completely neglected, but it does imply that foreign capital will have to be co-ordinated with the national and regional development programmes in such a way as to maximize its contribution to Spanish America's social and economic development. That is, the role of foreign capital—its potential contribution as well as its limitations—will have to be clearly understood by each of the member countries so that a consensus can be reached at the regional level as to the appropriate manner in which to treat it. This does not imply that *all* foreign capital in *all* member countries will be treated in a uniform manner. On the contrary, it is important that one make distinctions and discriminate in the treatment of foreign capital, but it also is essential that aggrement be reached as to the *principles* which will be applied in the regulation of this capital.

As an example, let us consider direct private foreign investments. This category includes investments which are financed, established, and operated by private foreign businessmen in the host country with the object of producing goods or services for sale (i) in the world market, (ii) in the integrated regional market, or (iii) in the national market of the host country. The nature of the market for which the product of the foreign investment is destined is very important, as different policies will be needed in each case to control the amount and location of this capital.

(i) Foreign investment for exportation to the world market is undertaken in three areas: plantation crops (bananas, cocoa, coffee), mining industries (copper, tin, iron ore), and petroleum. These industries often are organized, implicitly or explicitly, as cartels. This means that the producers face negatively sloped demand curves and, hence, invest, more or less, a fixed sum in order to obtain the volume of production necessary to maximize long-run cartel profits. Under such conditions the total volume of foreign investment in these primary commodities cannot, in general, be augmented. An individual country, e.g. by granting generous depreciation allowances, can increase its share of this type of investment *only at the expense of another country*. This tends to introduce competition between primary producers for the fixed flow of investment funds; this competition may take the form of provision for rapid amortization, the granting of special privileges, and the reduction of tax rates. The effect of this totally useless competition is to reduce the receipts of primary producers from foreign companies. If competing primary producers had a uniform policy with respect to foreign investment, *average receipts from foreign firms would be greater*; the average tax rate, for example, could be higher.

Even though LAFTA does not include all the competing primary producers it would be a distinct advantage if the member countries would co-ordinate their policies on foreign investment, and, in this way, eliminate much—if not all—of the wasteful competition to attract funds for the exploitation of mines, wells, and plantations.

(ii) Similarly, there is a need to co-ordinate policies on foreign investment when the subsequent production is destined for sale in the regional market. In this case both the total volume of investment as well as its location within the region are variable. That is, when production is destined for sale in the regional market there is no competition between LAFTA and third countries for a fixed flow of investment funds—as in case (i)—but there *is* competition *within* LAFTA over the location of this investment. Thus foreign investment which needs the entire regional market in order to be profitable can locate itself, presumably, in any one—or at least several—of the member countries, and which country, in fact, it chooses will depend in part on the level of restrictions in each member nation. Hence, once again, there will be a tendency for the member countries to compete among themselves for this capital by lowering restrictions. The benefits from foreign investment to the region as a whole will be much less than they would be if foreign investments were regulated regionally. In other words the region as a whole can increase its benefits from foreign investment if it acts together as a monopolist.[16] The difficulties of reaching such an agreement, of course, are obvious.

(iii) The final type of direct foreign private investment is that destined to supply goods to the national market. In this case there is little need for regional coordination of policies. Each nation should set its own level of restrictions so as to attract the volume and type of foreign investment it desires. It is to be hoped, of course, that the policies would be established so as to co-ordinate foreign investment with each national development plan. The point here, however, is that for this type of investment there is very little competition between the member countries: the fact that Paraguay is able to attract foreign businessmen to establish a cement plant does not preclude Ecuador from doing the same. The volume of foreign investment in each individual country is a variable but it is largely independent of policies in other countries.

Intra-union Factor Flows

In some circumstances intra-union factor flows and factor proportions may be important.[17] Four cases may be mentioned. First, a customs union between countries with high population densities presents no opportunity to gain from complementary factor proportions. These poor nations all are deficient in agricultural land and, in addition, are likely to have a meagre industrial and technical base; these nations are all deficient in the same factors. Second, a union between several small underdeveloped nations with 'favourable' land-labour ratios will result primarily in giving these economies the advantage of economies of scale. Factor flows in this case are likely to be unimportant. This is the case which in general applies to LAFTA.

Third, considerable potentialities for intra-union factor flows exist in a customs union between an over-populated area and an under-populated area. Here the factor endowments of labour and land are

[16] This argument is further developed in K. Griffin and R Hoffman, '*Planificación Nacional e Integración Economica Regional en América Latina*', in Instituto de Economia , *Simposio Integración Economica*, Santiago, Chile, 1965, pp. 37–41.

[17] See C. N. Vakil and P. R. Brahmananda, 'The Problems of Developing Countries', in E. A. G. Robinson, ed., *The Economic Consequences of the Size of Nations*.

complementary although scarce capital still will have to be created from domestic savings. The success of such a union will depend upon the willingness of labour unions in the labour scarce region to permit significant population movements into the area. There is a constant danger, which must be guarded against, that the emigration from the labour abundant region will consist primarily of young, energetic, and skilled adults. If this occurs the heavily populated region may suffer a decline in its rate of growth and the age distribution of its population will become more unfavourable. This process may be occurring in Greece, but such migration is unlikely to be important in Spanish America.

Finally, for the sake of completeness, we must mention the possibility of a union between a relatively capital abundant economy and a labour abundant economy. Such a union, however advantagous it might be, is not a realistic possibility in Spanish America. The Latin American Free Trade Area is a union of economies which are short of capital. Internal flows of capital would have no 'equilibrating' role to play and, to the extent that they are allowed to take place, they probably would be 'disequilibrating'—moving from the poorer to the less poor region.

Customs Unions as a Stimulus to Competition

Liberal economists tend to give great importance to the increase in competition resulting from a customs union. The invasion of protected national markets by aggressive outsiders is expected to energize the defensive, unenterprising, established monopolies. It is certainly true that in the short run a reduction or elimination of tariffs can be a powerful anti-monopoly device. Furthermore, tariff reductions may provide existing firms with a stimulus to innovate and expand. The formation of a customs union may even create favourable expectations of growth and thereby generate a self-sustaining dynamic process. It is not clear, however, that competition can be stimulated *only* by joining a customs union, or that joining a union is the *best* way of increasing competition.

Competition can be increased easily by unilateral tariff reductions. Thus if the primary benefit of integration is increased competition, this can be achieved without bothering to form a customs union. Perhaps an even better way to stimulate competition, if the domestic market is large enough, is through government established and managed firms which compete directly with private business. Such government corporations can prevent the undesirable practices of cartel agreements and at the same time provide the government with a direct means of control over investment and growth policies. Thus, where feasible, government owned firms are useful in two senses: they *discourage* anti-social monopolistic activity and *encourage* or facilitate expansion programmes. Particularly when economies of scale are important government owned enterprises may achieve greater efficiency than imperfect competition between private oligopolies.

The competitive struggle initiated by the formation of a customs union may be of relatively short duration. The sudden exogenous change may stir things up and lead businessmen to search for a new *modus vivendi*. That is, the competition may be only 'a transitional phase towards the establishment of monopolistic power'.[18] Thus the result of the union may not be increased competition but the appearance of extremely powerful international cartels. If the Regional Authority is less strong than the individual cartels (as is quite probable in Latin America) the power of locating industry and controlling their development will pass from national governments to private international cartels. For these reasons, the formation of a customs union is neither a necessary nor a sufficient condition for sustained competition.

Import Substitution at the Regional Level

If one of our small Spanish American nations is unable to sustain its growth on exports it may be forced to choose a process of import substitution based either on full capacity, sub-optimal (high cost) plants or upon optimum sized plants suffering from excess capacity. If the country builds plants on the basis of existing demand it will lose the advantage of economies of scale. If plants are constructed ahead of demand the country will lose the advantages of full capacity utilization. For this and other reasons the establishment of import replacing industries as a tool for economic development reaches an abrupt limit when the quantities of goods imported are no longer sufficient to justify the erection of a minimum sized efficient plant. Customs unions, by increasing the size of the protected market, widen the useful limits of this tool, and it is largely for this reason that Spanish America became interested in economic integration.

Before the process of import substitution began the economies of Spanish America were vulnerable primarily on the demand side. This was due, among other things, to the transmission abroad of business cycles originating in the advanced countries and to the low income elasticity of demand for the exports of the primary producing

[18] P. P. Streeten, *op. cit.*, p. 33.

countries. As Spanish America has come to rely excessively on import substitution, however, their economies have become vulnerable on the supply side. If imports are reduced to the bare minimum, any reduction in the capacity to import, for example due to an adverse shift in the terms of trade, will cut off the importation of essential items and this in turn may seriously damage the economy.

This is what appears to have happened in Spanish America in the last half of the 1950s. Export prices fell and, consequently, the terms of trade and capacity to import declined. This had two effects. First, government tax revenues from exports were reduced sharply, a deficit was created, inflation accelerated, and public investment was somewhat reduced. A decline in export demand, far from being deflationary, can be highly inflationary in fragmented economies of the type we are considering.

Secondly, imports of raw materials and intermediate goods were restricted in vain attempts to control the balance of payments. This forced existing industries to operate below capacity and further discouraged investment. As a result the industrial growth rate began to slacken. That is, a forced reduction in the supply of imported inputs disrupted industrial production, reduced the incentive to invest, and lowered the growth rate.

Again, the formation of a customs union can be helpful by giving a country an alternative to import substitution, *viz.* developing manufactured exports to other member countries and thereby increasing the capacity to import. This process essentially shifts the locus of import substitution from the national to the regional level. In this way the advantages of specialization can be preserved. At the same time the vulnerability of the economy to fluctuations originating in the wealthy nations can be reduced.

Discontinuities and Decreasing Costs

A nation's borders represent a discontinuity in the mobility of factors and goods.[19] It long has been recognized that the movement of commodities, labour, and capital is much greater within rather than between states. This is because currency and fiscal systems differ markedly between countries, as do customs, language, and education. Geography frequently helps to make a political and economic discontinuity a physical one as well. These 'natural' causes of discontinuities normally are reinforced by artificial causes: the movement of goods and factors is slowed further by tariffs and other trade restrictions plus the entire apparatus of the welfare state. Strictly

[19] See the introduction to E. A. G. Robinson, ed., *op. cit.*, p. xiv.

speaking, of course, the trade barriers due to differences of language or tradition are no more 'natural' causes of a discontinuity than is a quota scheme; they are historical phenomena. The important point for this discussion, however, is that the 'natural' barriers are 'given' data for the policy maker whereas 'artificial' barriers are subject to policy decisions and can be changed.

One of the important facts of contemporary economic relations is that the importance of the nation as a discontinuity is increasing. This is true for four reasons. First, the welfare state, with its emphasis on protecting the weak, is growing, primarily but not exclusively, in the industrialized countries. As most welfare benefits can be obtained only by resident citizens, this limits international mobility. Secondly, the number of independent states has grown rapidly, especially in Asia and Africa. The homogeneity of the colonial empire system is breaking down, and in its place a large number of discontinuous units has been created. Third, nationalism has become a strong political force in the new nations and by no means has vanished in the old. This emphasis on national differences rather than similarities accentuates discontinuities. Finally, the idea of economic planning and interference with the market mechanism is becoming more widespread; actual planning, however, has been confined largely to nation states. Hence, both as a cause and a consequence of the presence and growing importance of the nation as a discontinuity, economic planners have come to rely on the internal market as the chief vehicle for development. At the same time, this almost exclusive reliance on the domestic market has meant that smaller[20] countries in particular have had to forgo the advantages of economies of scale.

Many economists believe that the greatest benefits associated with customs unions in underdeveloped nations lie in the possibility of achieving economies of scale and thereby increasing industrial productivity. Their argument runs as follows: the larger is a country, the larger is likely to be its domestic market; the larger the domestic market, the larger the size of the firm it can support. Larger firms obtain lower unit costs of production through economies of scale. Thus a customs union, which has the effect of increasing the economic size of a nation, permits its member countries to exploit economies of scale and location by encouraging big firms to replace smaller and less efficient firms. Production costs are reduced; productivity is increased.

Stated in such simple terms it is not certain at all that the argument

[20] By small country we mean a nation that has a small population and gross national product plus a low *per capita* income.

is correct. The analogy between a customs union and a large nation surely is false. The removal of only some of the artificial causes of discontinuities, *viz.* tariffs, does not eliminate the natural causes nor the remaining artificial barriers. The national markets will continue not to be integrated so long as differences in traditions, languages, units of measure, or methods of distribution persist. Furthermore, as Professor Harry Johnson has pointed out, there is a qualitative difference between a mass of consumers and mass consumption.[21] It does not follow necessarily from the existence of a large market that tastes are sufficiently uniform to justify mass production techniques. Europeans are famous for their fastidiousness. Russians and North Americans, on the other hand, voice few objections to standardized commodities; neither, it seems, do Spanish Americans. Even if fastidiousness is important in the demand for consumer goods, it is a less important factor in the demand for producers' goods. The economies of scale that exist in these industries certainly can be achieved by widening the market.

Scale economies must be extremely important in Spanish America and most of the other underdeveloped economies. For example, Paraguay and Ecuador combined have a population only slightly larger than that of London and a *per capita* income that is much lower. They could take advantage of scale economies in many industries if their markets were bigger. Both the markets for final and intermediate products are too small to support one optimum sized plant in each industry. The problem is particularly serious for intermediate goods, since an economy which is unable to sustain an efficient plant in the production of final goods hardly could be expected to provide a large enough market to support intermediate inputs. Professor Jewkes has argued that this difficulty can be avoided if the smaller countries produce the intermediate products and export them to the giant firms in the larger nations. He presents figures showing that the 'United States Steel Corporation buys goods and services from 54,000 outside suppliers. The Ford Company of America in 1949 was reported to be purchasing 20,000 items costing round $700m from 7000 vendors.'[22] This suggestion, however, will be difficult to apply in practice. Most producers want their suppliers to be fairly near so that planning and consultation over technical details can be arranged easily. Furthermore, as argued in detail

[21] 'The Criteria of Economic Advantage', in G. D. N. Worswick, ed., *The Free Trade Proposals*, p. 32.

[22] 'Are the Economies of Scale Unlimited?' in E. A. G. Robinson, ed., *op. cit.*, p. 101.

below, these industrial complexes are highly interdependent industries; it is unlikely that they can be dispersed without influencing the profitability of the individual firms. In fact, in our opinion, the underdeveloped economies of Spanish America should try to develop entire industrial complexes and not just some of the component industries.

Summary

The potential gains from a customs union may be divided into two categories, those which improve the *allocation* of resources and those which lead to an increase in the *quantity* of available resources. The former are static, once-for-all gains while the latter are cumulative gains. The static benefits can be represented as derived from changed price relationships *along* a fixed production frontier and as a movement *to* the frontier. The dynamic benefits of customs unions, on the other hand, lead to an increase in the rate of economic growth, and can be represented best by a series of *shifts* of the production frontier. The potential increase in welfare from improved resource allocation is likely to be small. The benefits from dynamic processes, as we try to demonstrate below, may be somewhat larger. Whether these potential benefits are translated into actual gains will depend upon whether the member countries are able to overcome certain problems of short-run adjustment. There is no certainty they will be able to do this.

3 THE DISTRIBUTION OF BENEFITS

The restrictions governing international trade may be of a political or economic nature and the risks associated with them may be due to changes in tariffs, quotas, exchange rates, consumer tastes, or rates of inflation. The effect of these uncertainties is to make investment in export industries more risky than investment in corresponding domestic industries.[23]

A customs union can remove some of the uncertainty surrounding international economic relations, namely those associated with tariffs and other trade restrictions. As long as co-ordination of policies among the member countries is incomplete, risk and uncertainty originating from other areas will persist. In particular, uneven rates of inflation may make some countries' prices uncompetitive; unforeseen changes in internal monetary and fiscal policies will have external repercussions, etc.

[23] See B. Balassa, *The Theory of Economic Integration*, p. 177.

Such uncertainty as a union can eliminate depends upon the fact that membership is non-reversible. Once in a customs union, it must be made difficult for a country to leave again, for if membership is reversible uncertainty about trading relationships is not reduced and may even increase. Entrepreneurs will be reluctant to invest in industries the products of which will be exported to member countries unless they have some assurance the proposed trading relationships will persist for a reasonable period of time. What constitutes a reasonable period of time depends upon the depreciation rates of the newly established industries. Significant reductions in risk and uncertainty can come about only if irreversibility of membership is associated with close government co-operation and control by a regional authority.

The provision of an expansionary environment in which investment can take place is perhaps more important than the provision of savings to finance the capital formation. For if those who save also invest, i.e. if saving and investment decisions are not independent, but on the contrary, are interdependent, then the creation of investment opportunities may create the savings necessary to finance the capital accumulation. If the level of savings and investment is, in part, a function of investment opportunities, then planned investment co-ordination which distributes industrial complexes to each of the member states will raise the level of capital accumulation in the region and, hence, the aggregate rate of growth.

One thing is certain: the creation of a customs union will necessitate a series of adjustments within the member countries. All adjustments require investment to become effective. Hence the ability to invest and transform resources is an important criterion for the success of any customs union. If the capacity for adaptation of the member nations is low there is a great danger that the short-run problems of adjustment may become long-run structural problems. Customs unions bring certain changes, and from these changes the member countries are able to derive potential benefits. If, however, a country is unable to adapt to changed circumstances, the *potential benefits may become actual losses*. Thus the distribution of gains and losses within the union may be extremely unequal. The strong countries—those with relatively high *per capita* incomes, rates of growth, and capacity for transformation—may attract most of the gains, while the weaker members may be forced to bear the costs of adjustment. If the weaker members suddenly are forced to compete with their stronger partners, the effects might be disastrous. An inability to adjust may cause the destruction of important industries in the weaker nations: this will reduce the level of income absolutely. Furthermore, the risk and uncertainty created by strong competition may weaken the entrepreneurial class, reduce the incentive to invest, and thereby lower the trend rate of growth.

To avoid such a cumulative movement a series of measures will be needed. First, it is essential that the uniting countries distinguish between the weak and the strong,[24] for as Professor Streeten has said, 'The rules of equality do not apply to relations between unequals.'[25] The weaker countries must be granted certain safeguards and benefits to enable them to transform their economies to meet the new trading relationships. It is important, however, that the safeguards granted to the weaker members do *not* consist of the *negative* ones of tariff or quota schemes, for if the reimposition of restrictive controls is permitted no one will 'take the risk of planning investment on a scale which needs the whole of the area for profitable operation'.[26] Thus the weaker countries must adjust through expansion and not through contraction. This implies, secondly, the need for a relatively large investment fund administered by a regional organization. The purpose of such a fund would be to channel capital to the weaker members, the areas of slower growth. Bolivia, for example, has large forest reserves suitable for the production of cellulose. She also has undeveloped petroleum reserves and considerable potential in growing bananas. Even if the social overhead capital were available, which it is not, Bolivia could not develop these activities because she lacks the necessary venture capital. In this case the regional fund would have to provide long-term credit not only for the infrastructure but for the directly productive activities as well. Third, some agreement will have to be reached about the system of exchange rates to be followed by the union and the method of handling payments difficulties between member countries. Fluctuating exchange rates probably are not to be recommended. An unstable exchange rate will discourage investment that takes advantage of the regional economies of scale just as quickly as would tariff increases, other negative safeguards, or reversible membership arrangements. On the other hand, fixed exchange rates no doubt

[24] In Protocol No. 5 to the Treaty, LAFTA has recognized Paraguay and Bolivia as nations deserving special treatment in accordance with article 32. Ecuador later was given similar status. (The complete text of the Treaty together with the first five Protocols can be found in *The Economic Bulletin for Latin America*, Vol. V, No. 1, March 1960.)

[25] *Op. cit.*, p. 49.

[26] T. Balogh, 'Liberalization or Constructive Organization', in G. D. N. Worswick, ed., *op. cit.*, p. 45.

will lead to balance of payments problems, especially if rates of inflation among the member countries are unequal—as they are in Spanish America. Thus a regional Central Bank with a built-in mechanism and authority to provide short-term credit would be very useful.[27]

A customs union is in essence a trade liberalizing institution. A movement from national autarky toward free trade increases the danger that economic instability will increase while the measures to combat it will decrease. Some countries may encounter pronounced balance of payments problems and be left without any corrective remedy except that of deflation. In this way downward cumulative movements may originate. To avoid this problem and others associated with an inability to transform resources, a backward educational system, or poor social structure and mobility a series of *positive* safeguards will be needed within each country, but especially within the weaker countries. These safeguards should consist of specific controls over the type and location of investment, the right to subsidize new and important industries, and exchange controls necessary to avoid speculative capital movements and ensure that factors flow on the basis of complementarities, and not on the basis of cumulative movements. If these national safeguards are supplemented by a Regional Central Bank charged with avoiding deflation, and an Investment Fund responsible for ensuring sustained capital formation, plus a Regional Authority to co-ordinate the investment programmes and domestic policies of the member nations, there is a chance that an economic union could bring significant economic benefits to the individual participating countries. If, however, liberalization is not accompanied by co-ordination, there is no assurance whatever that a simple reduction in tariffs would bring lasting benefits.

4 THE TREATY OF MONTEVIDEO

The Latin American Free Trade Area was established on February 18, 1960, with the signing of the Treaty of Montevideo by Argentina, Brazil, Chile, Mexico, Paraguay, Peru and Uruguay. The following year Ecuador and Colombia joined the Association; Venezuela

[27] See R. Triffin, '*Una Camara de Compensacion y Union de Pagos Latinoamericana*', Centro de Estudios Monetarios Latinoamericano, *Cooperacion financiera en America Latina*, Mexico, 1963, pp. 95–117; ECLA, *Documentos sobre los problemas financieros, preparados por la Secretaria de la Comision Economica para America Latina para la Asociacion Latinoamericana de Libre Comercio*, Caracas, March 1961 (E/CN. 12/569), mimeo.

followed suit in 1967 and Bolivia immediately thereafter. Thus there are now eleven member countries.

Basic Provisions of the Treaty

The Treaty provides for the elimination of tariffs and other trade restrictions 'in respect of substantially all' of the reciprocal trade of the member countries.[28] This is to be accomplished over a twelve-year period,[29] although in practice the member countries have so far been totally unable to keep to the time schedule of the Treaty. Tariff reductions are intended to be affected on two schedules. The Common Schedule is a list of products whose tariff rates, by agreement, will be reduced to zero in all member countries. At the end of the first three years this schedule had to include at least 25 per cent of the intra-zonal trade. This stage was passed in 1964, but only after very intense negotiations in Bogotá among the member countries. In the subsequent three year periods the proportions of free intra-zonal trade were intended to increase to 50 and then to 75 per cent, and finally to 'substantially all of such trade' by the end of the twelve years.[30] In practice, the machinery has collapsed. As of 1968 negotiations to liberalize half of the region's intra-zonal trade—which could easily be done simply by including wheat and petroleum on the Common Schedule—have been suspended.[31]

The National Schedules, on the other hand, include products on which an individual nation grants concessions in an annual negotiating session.[32] This Schedule is constructed after each country compares lists of requests for concessions from the other member nations with its own list showing commodities in which the country is willing to grant preferences. Tariffs on products on the National Schedule must be reduced at least eight per cent a year although provision is made for withdrawing commodities from this Schedule if necessary.[33] There is no assurance, however, that products on the National

[28] See Article 3. [29] See Article 2.

[30] See Article 7. A formula for calculating the reductions in the weighted average of tariffs was established in Protocol No. 1.

[31] It should be noted, as R. F. Mikesell has written, that 'member countries only need to reduce their restrictions on the commodities that they are already importing from other members in substantial volume. Since this trade . . . is limited very largely to agricultural commodities . . . and non-agricultural primary materials and fuels, the required reduction in the average tariff can be brought about without the admission of any new products. . . .' ('The Movement Toward Regional Trading Groups in Latin America', in A. O. Hirschman, ed., *Latin American Issues*, p. 136.)

[32] See Article 5. [33] See Article 8.

Schedules ever will move to the Common Schedule, nor is there any guarantee that the concessions on these products someday will not be removed. Such an arrangement, as we argue in detail below, is unlikely to reduce the risk of regional investment or accelerate the rate of growth of the member countries.

Articles 16 and 17 of the Treaty state that the member nations 'may negotiate mutual agreements on complementarity by industrial sectors'. So far, however, little has been done to take advantage of this provision to co-ordinate industrial investment. The first agreement concerned data processing equipment and the second covered radio and television valves. The initiative for both came from foreign enterprises. Argentina, Brazil and Chile are considering a project in which each of the three nations would specialize in producing one or more components for the automobile industry, but a definite agreement has not been reached. In any case, it would appear—as argued in detail below—that the way to achieve manufacturing efficiency in Spanish America is not to break up an industrial complex and scatter its components over vast areas throughout the member countries, but to swap complexes, i.e. to let each nation concentrate on a single complex for export to its partners.

Escape Clauses and Special Measures for Weaker Members

The Treaty permits member countries to authorize another to impose non-discriminatory trade restrictions on a temporary basis if a large flow of imports is causing serious damage to an industry of 'vital importance to the national economy'.[34] Temporary restrictions for balance of payments purposes may also be imposed.[35]

Additional provisions are made to assist the development of the more backward members of the Association.[36] The weaker members may be granted special concessions by members which do not apply to all the others. The weaker members are authorized to eliminate their trade barriers over a longer period. Finally, the stronger countries are encouraged to provide technical and financial assistance to the weaker partners. As mentioned above, these measures originally applied to Paraguay, Ecuador and Bolivia; they have recently been extended to Uruguay. At the end of the Tercer Periodo de Sesiones in 1963 a new category of 'intermediate countries' was created. The countries in this category—Colombia, Chile, Venezuela and Peru—are to enjoy some preferential treatment, as yet unspecified.

[34] See Article 23. [35] See Article 24. [36] See Article 32.

The Weaknesses of LAFTA

There are many reasons why the Free Trade Association has been rather unsuccessful and why success in the future is somewhat doubtful. First, LAFTA, as its name implies, is a free trade area. This means that it suffers from all the disadvantages of such an arrangement and at the same time fails to enjoy the benefits of closer integration. In particular, most of the consideration discussed in Part 3 of this chapter in practice have been ignored.

Secondly, at present only ten per cent of the region's international trade is intra-zonal, i.e. only 1·5 per cent of the region's total income is traded among the member nations. Moreover, nearly 70 per cent of the intra-LAFTA trade is accounted for by Argentina, Chile and Brazil. The historically low volume of intra-zonal trade can be explained in part as the product of political and economic colonialism, in part as due to the absence of adequate regional transport facilities and in part as due to the autarkic policies of development pursued until recently by Spanish America.[37]

It was hoped, of course, that the value of trade between the member countries would increase greatly. There is little doubt that the potential for trade within LAFTA is very large, provided the region becomes industrialized. At the moment there is little basis for trade since all of the countries are essentially primary producers, but if the region is successful in accelerating industrialization intra-LAFTA trade could become more significant. Unfortunately, the potential for intra-LAFTA trade has not been exploited and, in fact, this trade has remained completely stagnant in the last few years: in 1965 the value of intra-LAFTA exports was $723·3 million; in 1967 it was still only $724·8 million.

The third major problem is the general lack of interest in economic integration outside intellectual circles. This problem exists at the international, region, and national levels.

The attitude of the United States towards regional co-operation in Latin America is quite distinct from its attitude towards European integration. After the sterling crisis of 1947 the United States channelled aid through the Organization for European Economic Co-operation and encouraged the common planning of investment among the member nations. This policy, in essence, has remained unchanged: the US continues to support closer economic and political ties in Western Europe as part of its campaign against com-

[37] Data on intra-LAFTA trade are available in D. W. Baerresen, M. Carnoy and J. Grunwald, *Latin American Trade Patterns.*

munism. In Latin America, on the other hand, the US seems to have remained apathetic to regional integration—in spite of the Declaration at Punta del Este and the April 1967 Meeting of American Chiefs of State in Uruguay. It appears that the United States prefers to deal with Brazil and Spanish America on a bilateral basis and fears, probably correctly, that close regional integration would diminish its influence in the area.

The International Monetary Fund, the International Bank for Reconstruction and Development and the Export-Import Bank have all been hostile or indifferent to close co-operation among the Spanish American nations.[38] Even the UN Economic Commission, which played such an important role in encouraging the establishment of LAFTA, appears to have lost some of its enthusiasm for the Association. Among the international agencies, only the Inter-American Development Bank is enthusiastic about promoting regional integration, and so far it has not been able to find many 'integration projects' to finance.

Enthusiasm within LAFTA itself does not appear to be very great. This is particularly noticeable in the Executive Committee of the Association: the budget is inadequate, no important studies of regional problems have been undertaken and a well trained civil service has not been recruited. At the April 1967 meeting referred to above, it was agreed to begin steps to form a Latin American Common Market linking LAFTA and the Central American Common Market. The initial steps are not planned until 1970, well after most of those who signed the document will be out of political office, and the formation of the Common Market is expected to take a further fifteen years. One certainly cannot accuse the LAFTA nations of blindly rushing forward into ill-considered actions!

Similarly, the Free Trade Area is failing to generate enthusiasm within each of the individual member countries. LAFTA suffers from a lack of public understanding and support and, with the notable exception of President Frei in Chile, it also lacks governmental enthusiasm and initiative. The three largest countries—Argentina, Brazil and Mexico—are anxious to maintain the *status quo* and are unwilling to make concessions which might strengthen the manufac-

[38] For a detailed discussion of the attitude of the IMF and the US toward Latin American integration see M. Wionczek, 'The Montevideo Treaty and Latin American Economic Integration', *Banca Nazionale del Lavoro Quarterly Review*, No. 57, June 1961. The essay is expanded and reprinted as 'La Historia del Tratado de Montevideo', in M. Wionczek, ed., *Integracion de America Latina*, Fondo de Cultura Economica, Mexico, 1964.

turing sectors of the other member countries. The latter, in turn, do not wish to be subjected to strong competition from the three large industrial nations and several of them are now forming a more closely knit group which would exclude the three big nations.

Just how successful will be the so-called Andean group of Chile, Bolivia, Peru, Ecuador and Colombia remains to be seen. The group hopes, of course, that their common interest in industrialization will enable them to promote policies for mutual advantage and to this end an Andean Development Corporation has been established. It is still much too early to judge the success of this initiative, but the fact that an attempt is being made to form a subregional association is indicative of the difficulties LAFTA has encountered.

In short, LAFTA has lost momentum due to the failure of the public and the government to appreciate the potential benefits of regional integration. Even in small countries like Uruguay, Venezuela and Paraguay vested interests fear the changes economic integration will create. They view the probability of loss greater than the probability of gain and, for this reason, oppose or are indifferent to LAFTA. Those who *actually* are going to lose are more vocal and powerful than those who *potentially* may gain. Thus public opinion appears to be apathetic or even hostile to the Free Trade Area. Furthermore, the political will is lacking. The member countries are unwilling to co-ordinate their development plans, reduce their sovereignty and organize investment on the basis of a potentially larger market. These problems are aggravated by the political instability of Spanish America. The government of Ecuador has oscillated between a constitutionalist and an extreme right wing military regime; Paraguay continues with its dictatorship; Argentina and Peru periodically have military rulers; only the democratically elected government in Chile is keen to accelerate integration. Bolivia is always on the verge of political turmoil and until recently Venezuela was plagued by organized terrorism. To establish an economic union between such societies is indeed a difficult task.

5 REGIONAL PLANNING FOR INDUSTRIALIZATION

Although the difficulties of accelerating integration in Spanish America are formidable, the potential gains are worth the effort involved. In this final section we propose to summarize the source of benefit of regional economic integration, to demonstrate the need for regional investment co-ordination and tentatively to suggest in general the appropriate criteria for sectoral investment allocation.

The Source of Benefit of Integration

According to neo-classical economic theory the benefits of a customs union are to be assessed primarily in terms of whether trade is 'created', by shifting sources of supply from high to low cost areas, or 'diverted', by shifting sources of supply from non-members to member countries. According to this static allocation criterion the Free Trade Area is a great mistake. Not only is trade diversion likely to be more important than trade creation, but the whole purpose of LAFTA seems to be the diversion and encouragement of intra-regional trade in industrial products. Article 28 excludes agricultural goods from most of the provisions of the Treaty, while Article 16 specifically encourages 'mutual agreements on complementarity by industrial sectors'. Thus, by its own inner logic, the 'success' of LAFTA is to be measured by its ability to displace trade in manufactured goods from low cost European and North American sources to the high cost sources within the 'club'.

The ultra-liberal economist[39] must believe the Treaty of Montevideo was signed in a moment of madness, for seldom have the tenets of 'sound' economics been so thoroughly disregarded; and yet, there is a method to the madness. From the point of view of the region's economic development it is *potential* not actual costs of production that are relevant. It is hoped, and believed, that economic integration will enable the present member countries *each* to develop high-productivity industrial sectors which, eventually, will be able to compete on equal terms with the products of the industrial nations.

The Investment Process in Spanish America

High productivity industrial sectors are not likely to arise if exclusive reliance is placed on the unguided price system as the mechanism of expansion. This is true, first, because the contemplated *changes* in the productive apparatus of the member countries are not 'marginal', but on the contrary, are 'structural'. In underdeveloped economies most economic policy decisions and all major investments (such as the construction of a steel mill in Concepcion, Chile) are concerned

[39] See J. E. Meade, *The Theory of International Economic Policy*, Vol. II. *Trade and Welfare*, Ch. XXXII; H. Johnson, *Money, Trade and Economic Growth*, Ch. III plus the appendix; (Chapter III was presented in Pakistan as a lecture; the appendix was published originally in India); J. Viner, *The Customs Union Issue;* J. Viner, *International Trade and Economic Development*.

with, and bring about, substantial changes in the functioning of an economy. Thus the *ceteris paribus* assumption of the traditional theory of prices and of marginal analysis in general does not apply.

Secondly, the *investment programme* under such conditions is one solid, interdependent 'lump'. That is, the profitability of one investment project will be influenced greatly by the absence or existence of a second project. Hence the crucial planning decision is not so much the evaluation of individual projects but the selection of *groups* of interdependent projects which, when taken as a whole, are mutually reinforcing. Third, the nature of the *investment process* in the less developed economies is quite distinct from that in industrial economies.

In the less developed countries investments by private entrepreneurs on balance tend to be competitive in the sense that the creation of a new production unit tends to lead to the proliferation of many similar establishments producing substitute goods. Thus the establishment of one successful textile mill may encourage other entrepreneurs to establish a series of similar mills. This activity eventually will create excess capacity in the industry and reduce its profitability. At the same time investments which create excess capacity in textiles will not present investors with many profit opportunities in other industries. Thus the effect of competitive investment is to lower the return on capital and, eventually, to discourage further accumulation. This appears to be a fairly general characteristic of investment in light industry, i.e. the area where private initiative is most important.

Part of the explanation for substitute production and competitive investment in the backward economies lies in the fact that for a series of reasons their small and inexperienced entrepreneurial class is not accustomed to *creating* investment opportunities, but instead is inclined merely to *respond* to obvious profit situations. A further possible explanation lies in the fact that the low and stagnant incomes of these nations do not provide opportunities for economies of scale. Depending on the form the scale economies take, two things may occur. If the investment is very 'lumpy' private entrepreneurs, even if they can raise the finance, will be unwilling to assume the risk of failure due to the limited size of the domestic market. In this case, unless the government undertakes the investment or initiates a policy of balanced (self-reinforcing) expansion of supply and demand, investment will not be forthcoming. On the other hand, if the economies of scale do not materialize until a certain volume of production has been reached, many high cost, low output firms will appear.[40] This, as mentioned above, will lead to excess capacity, a reduction in the yield of capital, and a decline in investment.

A still further explanation lies in the scarcity of possibilities for complementary investment in these countries. Most specialize in agricultural or other primary commodities in which the 'linkage effects'[41] are slight when compared with industrial production. In the language of input-output analysis, most of the cells in a production matrix would be filled with zeros or numbers which were not statistically significant. This is especially true in Asia and Africa. In the subsistence sector it is obvious that labour will form almost the only input. In cash-crop agriculture and in many other activities, whether production is for export or for domestic consumption, 'one has only to take a cursory glance at the structure of the industry to see that wages and profits almost entirely account for costs. . . . On the output side, export industries usually figure prominently in these countries and they obviously sell to one sector which is outside the production sector'.[42] Subsistence production, by definition, does not pass through a market, and cash-crop agriculture destined for the domestic market sells direct to the consumption sector.

Even in Spanish America—relatively advanced among the world's underdeveloped regions—a large proportion of the population lives in rural areas and over a third of the output is composed of agricultural or mineral products. If, for example, the price of coffee rises there is little that an investor, say, in Colombia, can do besides increase the production of coffee. He can't produce inputs for the industry because few inputs other than land and labour are used. That is, the few remaining inputs are imported goods the demand for which is insufficient to justify the erection of an import substituting industry. He cannot use the output of the industry as an input in a new venture because coffee, with few exceptions, is a final product. Thus the investor, and the economy, are 'trapped' in a rigid production structure which is difficult to escape in the absence of conscious planning. 'Backward linkages' are negligible, and the few 'forward linkages' that exist cannot be exploited because of insufficient demand.

Conditions are quite different in the rapidly growing industrial economies. Technical progress combined with rising income and

[40] An example of this is the production of refrigerators in Chile by many small and inefficient firms.

[41] See A. O. Hirschman, *The Strategy of Economic Development*.

[42] A. T. Peacock, and D. Dosser, 'Input-Output Analysis in an Under-developed Country: A Case Study', *Review of Economic Studies*, No. 66, October 1957, p. 22.

demand provides the physical and economic possibilities for further expansion. Technical progress constantly provides new inputs and outputs, i.e. it is continually creating new intermediate and final products. Growth and full employment generate an expanding market. Thus supply and demand conditions are changing continually in such a way as to create new profit possibilities. This leads entrepreneurs to look for new ways of providing inputs to new industries or of using the output of new industries as an input for a new product. Thus, on balance, the competitive effects of investment are outweighed by the complementary effects; bottlenecks and profit opportunities are created constantly, and the high profitability of investment helps generate a self-sustaining growth process.[43]

Integration and Industrial Complexes

It is (or should be) the primary purpose of industrialization in the Free Trade Area to accelerate growth through the provision of capital goods and to increase productivity through reductions in industrial costs. As long as productivity in the agricultural sector is less than that of industry, it would be possible to increase the productivity of the *economy* merely by shifting the composition of output toward industrial goods; no reduction in sectoral costs would be necessary. This does not seem to be the primary intention of those nations which signed the Treaty, however. The signatory nations want not only to change the composition of their output and exports, but especially, to increase their international competitiveness in industrial production. This can occur only by reducing manufacturing costs, and it is for this reason that Spanish America has become interested in regional integration.

The reductions in industrial costs mainly will come from two sources: the existence of factor indivisibilities and production interdependencies. First, the expansion of the market will enable some existing producers to increase output and reduce unit costs through *economies of scale*. Secondly, the preferential treatment granted to member nations will enable *infant industries, sectors, and economies* to 'learn by doing' and to develop. Thirdly, the creation of *industrial*

[43] In such an economic environment a certain degree of substitute production is highly desirable because it helps to impede the development of monopolies. Thus it is the *lack* of complementary investments, and not so much the excess of competitive investment, which in the underdeveloped countries is so unfortunate. One frequently observes in Spanish America that the few heavy industries are monopolies, entry being restricted by the significant economies of scale. Light industry, on the other hand, more often is characterized by a certain degree of competition.

complexes with their strong interdependencies on both the supply and demand sides will provide investment opportunities in other sectors. Hence the purpose of integration is to provide and expand several *growth points*, i.e. nucleii of industrial interdependencies, in the member countries. Such 'growth points' are not likely to develop rapidly or spontaneously, however. Their encouragement will require national planning and regional co-ordination. Unfortunately, none of the member states has begun seriously to plan economic development, although all have expressed intentions of doing so. Spanish American governments so far have contented themselves with making projections; no plan implementing measures are provided or discussed. Thus planning documents are little more than a list of hopes and aspirations—a book of prayers.

Even more discouraging, the commodity by commodity approach to tariff reduction, as apparently envisaged in Chapter II of the Treaty and as implemented in practice, is practically the antithesis of effective regional co-ordination. According to Article 18, if an individual country grants a concession on a particular commodity, it must do so for *all* the member countries. This is what is meant by 'most favoured nation treatment'. This does not imply, however, that any other country belonging to the Area must grant a similar concession on the *same* product. Thus there is no guarantee in the Treaty that a homogeneous Latin American market ever will develop in new industrial products.[44] For example, Ecuador may grant Chile (and hence all other member countries) a concession in steel tubes, but this does not mean that the other member countries also will reduce their tariffs on steel tubing. If there are important economies of scale in the production of this product the combined Chilean-Ecuadorian market may not be sufficiently large to justify the erection of a minimum sized efficient plant and thus the 'concession' will be meaningless. If, as we believe, the advantages of integration stem in large part from the greater market of the region as a whole, then LAFTA will be successful only if the member states *as a group* agree to tariff reductions. The crucial point is not whether the *importing* nation discriminates between member countries but whether the *exporting* nation can take advantage of the potential benefits of a unified Latin American market. If, in fact, a protected and fairly homogeneous Latin American market is not created, the hope of 'diversifying interchange' (article 10) and realizing 'the

[44] This matter is further complicated by the absence of a common tariff against non-member nations. As noted above, the member nations do not intend to begin the formation of a common market until 1970.

expectation of increasing currents of trade' (article 13) will be frustrated—as, in fact, has occurred so far.

Furthermore, as long as the negotiating sessions are concerned with reducing tariffs on, say, peaches and ashtrays, it is extremely unlikely that the formation of the Area will help develop important iron and steel complexes, an atomic energy centre, a petro-chemical complex, or a modern electronics, aluminium, electrochemical, plastics, or paper industry. This does not imply that any single country, through national planning, cannot develop a particular industrial complex. Tariff reductions and other concessions can be negotiated on a commodity by commodity basis provided the requests of the bargaining country are concentrated in a few, limited, lines of production, which are or could be strategic for its internal development.

There is no reason why government technicians could not select three or four industrial complexes which Chile, for example, is eager to develop over the next thirty years, and then ask the existing national producers to indicate the specific commodities within these complexes on which they would like immediate tariff reductions. There once were rumours that something of this sort had been happening in practice. Within the list of over 500 products on which Chile requested tariff reduction, it was claimed that the negotiators concentrated on steel, up-grading copper exports, wood products, and automobile components. Two things may be said about this rumour. First, a cursory glance at the 'petition list' does not convince one that there was a concentration on any particular product or line of production. Of course, the *bargaining* may have concentrated on only a very few products in a very diverse list. The results of the negotiations to date, however, do not support this hypothesis. Secondly, two of the product groups on which the negotiators supposedly are concentrating, *viz.* up-graded copper exports and automobile components, are not emphasized at all in either the so-called Ten Year Plan or its successor, the current three year Development Programme, 1968–70. It appears, then, that neither the 'integrationists' nor the 'planners' in Chile take into account what the other group is doing. In fact, it appears that so far there has been little policy co-ordination, little national economic planning, and little regional integration.

Cartel Planning

In the absence of *public* planning, the only remaining hope for establishing manufacturing complexes is through *private* planning, i.e. market and production allocation on the basis of industrial cartel agreements. Atomistic competition in heavy industry, far from being the unmixed blessing of liberal economics, would result in a slower rate of growth through a failure of investment co-ordination.

Planning through cartels might be an effective way to achieve growth in Continental Europe, but it is unlikely to prove effective in Spanish America. Two reasons account for this. First, a necessary condition for the development of industrial complexes is the *prior or simultaneous existence* of social overhead capital. Such social capital is lacking or notably insufficient in Spanish America. In particular, energy and transportation facilities are inadequate, nor is the educational system geared to producing skilled labour and trained technicians. The latter is an important point, for the essential characteristic of many modern industrial complexes is that their more important inputs are knowledge and specialized human resources. A country's 'comparative advantage' in these industries is not 'given' by its natural endowments but is *acquired* either through historical accident (e.g. immigration) or planned development (e.g. Russian missiles). A striking feature of the United States is the attraction exerted by, and the concentration of industrial complexes around, the intellectual centres—universities, technical schools, and research institutions. These organizations provide the crucial inputs for the industrialization process. The less prosperous regions of the country are precisely those where the education system is backward. Similarly, the less prosperous regions of the world are those where cultural values and the educational hierarchy are oriented in non-scientific directions. A necessary condition for the industrialization of Spanish America is the creation of an adequate, scientifically oriented educational system.

Secondly, even if private cartels could overcome the handicap of inadequate social overhead capital and a backward educational system, sustained growth is unlikely. This is true, not so much because monopolies are inefficient, but because cartel planning is likely to result in extremely unbalanced regional growth. Investment will tend to be polarized around the present manufacturing centres.[45]

[45] J. Tinbergen makes a similar point: 'Unlike most agricultural and mining activities, industrial activities can be carried out in many places, provided only that the transportation facilities are adequate. Where to begin is a matter of relative indifference to the private entrepreneur. Once he has started, moreover the external effects and the infrastructure created will put a premium on new investments in the same centre. These combined forces make the creation of centres of heavy industry under free enterprise a somewhat haphazard process

The relatively more backward regions will remain backward and will serve only as a market for the industrialized zones. The lagging countries may soon withdraw from the area, unless a compensation mechanism is created. Neither the escape clauses of Chapter VI of the Treaty nor the special measures in favour of the weaker members (Article 32) are adequate to prevent the backward economies from regressing further. This is because industrial competition between economies of unequal strength and resiliency is likely to result in the destruction of the industries of the weaker member, and not merely in a reallocation of factors of production. A strong competitive blow to an important industry in a weak country 'will not merely necessitate a once-for-all adjustment but might leave the weaker part of the area absolutely poorer without hope of recouping itself unless very strong measures are taken to boost investment. The uncertainty and risk caused by the existence of strong and powerful competition within the combined area will reduce investment and thus retard and impede the change in the productive structure which is needed if the weaker area is to recover. . . .'[46] Thus the probabilities are fairly high that if investment projects are not co-ordinated at the regional level the weaker members will invoke Article 64 and withdraw from the Area,[47] or a new sub-regional group will be formed.

As LAFTA is presently organized those who will gain the most from integration include the strong industrial producers in Mexico, Brazil and Argentina along with the powerful foreign enterprises and the financial and commercial interests linked to them. The small industrialists and artisans of Spanish America are likely to be subjected to intense competition from the larger producers located in favourably endowed regions. Rather than passively accept an absolute decline in their material wellbeing they may mobilize political support and

Some centres may in the end become too large from a social point of view, while in other very desirable locations they may not be set up at all. Market forces are no longer a clear and unambiguous guide to what is the most desirable distribution of heavy industry over the area under consideration'. (See 'Heavy Industry in the Latin American Common Market', *Economic Bulletin for Latin America*, Vol. V, No. 1., pp. 3–4.)

[46] T. Balogh, 'Unequal Partners', in G. N. D. Worswick, ed., *The Free Trade Proposals*, p. 129.

[47] This problem cannot be solved by extending beyond the present five years the period in which the withdrawing nation is required to maintain its previous concessions. First, the article probably is not enforceable, and secondly, the interests of the Area lie not only in preventing the present members from leaving but in broadening the association to include the Central American Common Market countries.

attempt to withdraw, possibly forming a more congenial association such as the Andean Group.

Regional Growth and the Distribution of Investment

If the weaker members withdraw, the dynamic regions will lose part of their markets; some of the opportunities for taking advantage of the economies of scale will disappear, or at the very least, will be substantially reduced; and the growth of these areas will be retarded. In fact, the mere possibility that some nations may leave LAFTA is enough to ensure that no entrepreneur will assume the risk of investing on a scale which needs the entire area for profitable operations. That is, the greater the risk that LAFTA may shrink, the smaller will be the newly constructed plants. When one considers the small population of most member countries, the low *per capita* income, and its unequal distribution, it is probable that *nearly every* major industrial complex will need *at least* the whole of LAFTA for profitable operation. There are significant economies of scale in almost every durable product, e.g. aircraft, ship building, locomotives, railroad and streetcars, tractors, motorcycles, rayon, typewriters, glass containers, and office machinery.[48] The minimum and optimum size of plant may be rising along with technical progress, and furthermore, there is some evidence that cost curves may be continuously falling, i.e. the economies of scale may be unlimited.[49] Equally important, a market 'that is large enough to provide adequate . . . outlets for the output of at least one optimum-sized plant in all industries producing final products may still be sub-optimal if some of these plants need equipment, servicing, or other intermediate products, but provide too small a market outlet for some of these'.[50] It is quite probable that the withdrawal of some members from the Free Trade Area would entail economic losses for the rest. Hence *each* country has an interest in ensuring that every other country receives some of the benefits of integration. This can be achieved only by regional investment co-ordination.

[48] See J. Bain, *Industrial Organization;* P. Sargent-Florence, *Post-War Investment Location and Size of Plant;* H. B. Chenery, 'Patterns of Industrial Growth', *American Economic Review*, September 1960; F. T. Moore, 'Economies of Scale: Some Statistical Evidence', *Quarterly Journal of Economics*, May 1959.

[49] See S. Hymer, and P. Pashigian, 'Firm Size and Rate of Growth', *The Journal of Political Economy*, May 1959.

[50] T. Scitovsky, 'International Trade and Economic Integration as a Means of Overcoming the Disadvantages of a Small Nation', in E. A. G. Robinson, ed., *op. cit.*, p. 283.

The Polarization of Investment: Historical Examples

The polarization of investment and the impoverishment of the backward regions indeed has occurred historically. The union of Ireland and Britain (1800–1921) had disastrous effects on Irish industrial and agricultural production; starvation, pestilence, and the widespread emigration associated with the Great Famine caused a catastrophic decline of population. The formation of Great Britain resulted at least in the relative impoverishment of Wales and Scotland. The creation of unified Italy wrought the destruction of southern Italian industry. A great Civil War was fought in the United States to prevent the Southern agricultural states from leaving the industrial North. These customs unions did not dissolve, even though the benefits of integration were shared unevenly, because in addition to being economic unions, they were strong political units as well.[51] Ultimately, as in the United States, *force* could be used to hold the union together. More often, the effects of an unequal distribution of investment were compensated, in part, through direct state controls and a national fiscal policy which taxed the wealthy regions and gave to the poor.

This compensation mechanism might be viewed as an application of the well known Kaldor-Hicks test of welfare economics. According to this test, a contemplated change is potentially desirable if those who would gain from the change could afford to over-compensate those who would lose. Subsequent discussion amended the test to read as follows: The change in practice is desirable only if those who lose are compensated *in fact* or if it is deemed the distribution of income has improved. This is one of the roles of progressive taxation and government expenditure in a modern state, *viz.* to improve the distribution of income and to compensate those regions and persons who are harmed by economic change.

The Need for Investment Co-ordination

The Free Trade Area, however, is not a unified state. It has no federal fiscal policy which can channel income to the backward areas. Nor can it force private investors, either foreign[52] or native,

[51] An exception to this generalization is the withdrawal of the Irish Free State from the United Kingdom in 1922.

[52] In the case of foreign investments, it is probable that these investments will be attracted to those markets which already are large and growing. Thus capital imports may further accentuate regional polarization by increasing income inequalities and the differences in the rates of growth.

to invest in projects or areas not of their choosing. Thus it is essential that *public* investment funds be created and allocated in such a way as to initiate and maintain balanced regional growth. That is, at the regional level growth must be balanced among nations in the sense that no nation should lag too far behind; at the national level, growth need not be balanced among regions. Each country must have at least one industrial complex so as to distribute the benefits of integration and ensure the maintenance of the association. The complexes, however, cannot and should not be distributed evenly (like butter on bread) throughout each country. On the contrary, it is expected that areas of intense science-based industrial activity will coagulate, perhaps around the intellectual centres. The creation of these growth points is essential if full advantage of externalities is to be taken.

The question still remains as to precisely what planning methodology should be used to allocate the industrial complexes among the member countries.[53] Let us assume

(a) the planners are interested primarily in *industrial* investment, not only because this is the evident concern of the signatory nations as indicated in the Montevideo Treaty, but also because this is the most promising area for future advances in productivity and output;

(b) the planners are interested in *new* industries and for practical and political reasons they want to avoid saying, for example, that existing industry *x* in country *y* should be contracted because it is relatively less efficient than the same industry in another member country. This would exclude such things as the steel industry and automobile assembly plants from the planners' concern, and would focus attention on establishing industries which can substitute products presently imported from the industrial nations, particularly capital goods.

(c) The planners are interested in industries characterized by significant *economies of scale*, i.e. industries which will need the entire Latin American Market for efficient operation or which at least require markets larger than those existing in any one single country. If economies of scale are unimportant then, as argued above, the whole *raison d'être* of Latin American integration collapses.

If the above assumptions are accepted the planners will have to

[53] W. Baer and I. Kerstenetzky ('Import Substitution and Industrialization in Brazil', *American Economic Review*, May 1964) have analysed (*ex post*) the recent development of Brazil using the same techniques recommended for *ex ante* planning of industry in LAFTA. The present distribution of industrial centres is analysed in P. R. Odell, 'Economic Integration and Spatial Patterns of Economic Development in Latin America', *Journal of Common Market Studies*, March 1968.

focus their attention on so-called 'heavy' industry. One of the major features of 'heavy' industry, aside from the scale factor, is the high degree of complementarity between projects within a given industrial complex. The consequence of this interdependency is that the profitability of investment in any of the component industries is greatly influenced by the existence of the other elements of the complex. In general the complementarities and externalities may arise from one of two sources:

(i) the common use among several firms and industries of an important raw material.[54] These complexes will be called 'raw material-based complexes'. They are similar to Hirschman's 'forward linkages'[55] and in a table of inter-industry sales and purchases they can be isolated by reading *down* the columns. Typical examples of raw material-based complexes would be coal mining or the chlorine-caustic soda complex;

(ii) the purchase by a single industry of a large proportion of the output of its principal suppliers. These complexes will be called 'final product-based complexes'. They are similar to Hirschman's 'backward linkages' and can be isolated by reading *across* the rows of an input-output table. Shipbuilding is a typical example of a final product-based complex.

The most important methodological problem is to define precisely what is meant by an 'industrial complex'. Perhaps the best way would be to define complexes empirically using input-ouput data as discussed above.[56] Provided the degree of aggregation is not too great this should enable the planners to spot the important inter-industry flows and, presumably, to isolate the industries in which complementarities predominate. The required data certainly are available for the United States, France and Britain and probably are available for several other Central and Western European countries.

In using inter-industry data from presently industrialized nations as a guide to isolating industrial complexes we are implicitly assuming that Spanish America should use the most advanced techniques of production even though factor proportions differ quite markedly between the two categories of countries. The misallocation of

[54] A special case of this would be the common use in firms (in which inter-industry flows may be small) of similar types of skilled labour.

[55] See A. O. Hirschmann, *op. cit.*, Chapter VI.

[56] Alternatively one could work with an 'index of the sensitivity of dispersion' and with 'key industries' as determined by an 'index of the power of dispersion'. These indexes are not unlike measures of forward and backward linkage. Cf. P. N. Rasmussen, *Studies in Inter-Sectoral Relations*, (Copenhagen, 1956), Chapter 8.

resources caused by accepting this assumption is likely to be small, however, as in many cases the range of alternative techniques will be small and, in cases where labour intensive methods are available, their use may lead to a reduction in the quality of output.

Once complementary relationships have been determined through inter-industry flows the planners must determine which, if any, of the component industries enjoy decreasing costs through economies of scale. They then must determine from among those groups of industries characterized both by important complementarities and economies of scale which *also* enjoy geographic proximity—for if there are no economies of conglomeration there is no need to establish 'growth points', i.e. to concentrate the entire complex in one member country.

Whether or not industrial complexes tend to be concentrated geographically can be ascertained by examining census data in the industrialized countries or by using atlases which include maps of industrial location. We thus assume (a) that the growth processes (either through market forces or planning) which historically have encouraged geographical concentration in industrial nations also will be operative within LAFTA and (b) that these processes were 'rational' in that they led to costs lower than would otherwise have prevailed.

Once the complexes have been isolated through the above procedure the planners will be ready to make cost estimates. At this stage, however, they will want to exclude from costs all elements which are purely historical phenomena. That is, they will not want to mechanically apply criteria for static resource allocation—which assume given factor endowments—in circumstances in which factor supplies can be increased by conscious decisions; we do not want a country to prejudice its possibilities for industrialization simply because it is not industrialized. This means that in comparing the cost of establishing a given complex in each member country the planners should assume the cost of labour and capital to be the same in all countries, even though, say, skilled labour may be relatively more abundant—and hence relatively cheaper—in Argentina than in Colombia. If one follows this procedure then the only cost differences which should be considered are raw material costs and potential[57] transport costs.

[57] In cases where transport facilities are inadequate or non-existent, the calculations should include an estimate of what transportation costs *would be* if appropriate investments were undertaken. A country should not be forced to remain non-industrialized simply because in the past it failed to develop adequate transport services.

When the cost estimates are finished a first ordering of the complexes by profitability can be established. This ordering then should be adjusted by considering the following additional factors:

1. The time necessary for the projects in the complex to reach maturity. The weighting element for this factor presumably should be the social discount rate, but since this is unknown it will have to appear implicitly as the result of political negotiation;
2. the desirability of diversifying national production;
3. considerations of income distribution within LAFTA. As stated above, each member nation should be allocated at least one industrial complex.

The funds necessary for the massive investment effort visualized could be raised in part from the member countries, for example by taxing foreign corporations, and in part from foreign aid institutions. As regards the latter, it would be desirable that a large proportion of the public foreign aid that now is given to the member countries individually be given to a regional investment co-ordinating agency which would be responsible for allocating these funds on the basis of the needs of LAFTA as a whole. Whether a new organization is created or the functions of the proposed agency are taken over by an existing institution, such as the Inter-American Development Bank, is relatively unimportant. What *is* important is that the member nations begin to think in terms of regional investment planning. Paradoxically, the *less* interested are the member nations in political unification the *more* necessary is regional investment co-ordination to ensure the union does not partially disintegrate. A simple Free Trade Area is not likely to stimulate Spanish America's economic development. Closer integration is essential if the weaker partners are to avoid impoverishment and the region as a whole is to overcome the disadvantages of backwardness and learn to compete with the industrialized nations.

CHAPTER VII

Spanish America and the Industrial West

We are approaching the end of our enquiry. The time has now come to gather the threads of our story together and consider the conclusions that emerge. Looking back, the story we have told is not a happy one.

The fundamental characteristic of Spanish America, of course, is its poverty and the enormous disparity in income between the industrial West and our nine republics. In 1965 the income of General Motors was $20·7 thousand million. No Spanish American country had a gross national product as large as this; indeed, the GNP of our largest country—Argentina—fell short by about 25 per cent. In 1968, according to *Fortune* magazine, there were six US citizens worth between $500 million and $1,000 million and another two worth between $1·0 and $1·5 thousand million.[1] The assets of the first group exceed the national income of Bolivia and Paraguay while the wealth of J. Paul Getty and Howard Hughes—the two richest Americans—is greater than the national income of Ecuador and nearly as large as that of Uruguay. In other words, the wealth of a few individuals and the sales revenue of a few large corporations exceed by a multiple the national income of most Spanish American nations. In 1966 the GNP of Spanish America as a whole was approximately $42·2 thousand million, i.e. little more than the annual income of two corporations the size of General Motors.

There are 153 Americans worth $100 million or more. The total wealth of this small group is approximately $36·7 thousand million. In 1966 the combined GNP of Paraguay, Bolivia, Ecuador, Uruguay, Peru and Chile was about $13 thousand million. Assuming a capital-output ratio of 2·8, this income was produced by a capital stock worth roughly $36·4 thousand million. That is, the 153 Americans were rich enough to purchase the entire stock of capital of these six

[1] *Fortune*, May 1968.

nations. No doubt such a take-over bid is not being considered, but our example illustrates the power which wealth confers.

Not only is the disparity in income between Spanish America and the industrial West very large, it is quickly becoming larger. Even if the rates of growth of *per capita* income were identical in the rich and poor nations the absolute disparity in income would increase because of the difference in base. For instance, consider two countries growing at 5 per cent a year per head, in one of which *per capita* income is $2,000 and in the other $400. The initial disparity, thus, is $1,600. But in the first country income will rise by $100 at the end of the year to $2,100, while in the second income will rise by only $20 to $420. The disparity, therefore, will increase by $80 to $1,680. In fact, in our example, given the rate of growth of the rich country, a widening absolute disparity could only have been prevented if the poor country had achieved a rate of growth of *per capita* income of 25 per cent!

Conditions in Spanish America are very much worse than this, however. Since 1957 the rate of growth of *per capita* income in Spanish America has been less than 1·5 per cent a year. In Europe, in contrast, the rate of growth has been nearly 4 per cent and even in the United States, which is generally considered to have grown slowly, the *per capita* growth rate has been nearly 2·5 per cent. No Spanish American economy grew as rapidly as Europe and only one—Peru—grew faster than the United States. Thus Spanish America as a whole has been losing ground to the industrial nations; not only are absolute income disparities widening, but the rate of growth and relative incomes are lagging behind.

In some Spanish American countries conditions have actually become worse for a large proportion of the population. This has certainly occurred in Uruguay and Argentina. In others, as we argued in Chapter I, the increments of *per capita* output are channelled to the wealthier individuals and areas, so that both the personal and the regional distribution of income become more unequal. Numerous studies, both theoretical and empirical, are now available which demonstrate that regional inequalities increase with development.[2] Indeed, the poor regions frequently are net contributors of resources to the rich regions.

Moreover, the distribution of personal wealth tends to be more unequal in the richer regions. A study of Peru by David Chaplin,

[2] See, for instance, J. G. Williamson, 'Regional Inequality and the Process of National Development', *Economic Development and Cultural Change*, July 1965; W. Baer, 'Regional Inequality and Economic Growth in Brazil', *Economic Development and Cultural Change*, April 1964.

for instance, finds that the distribution of wealth in the form of real estate increases 'as we move from the isolated mountainous departments to the commercialized plantations on the coast. Unequal as the distribution of land was in the interior it was more so on the sugar and cotton haciendas, especially on the north coast.'[3]

Per capita agricultural production, and particularly food production for domestic consumption, has—in many countries—failed to keep up with the rate of population increase. This is a major reason why in countries like Paraguay the rural poor are becoming further impoverished. For large numbers of people not only is the international distribution of income becoming more unequal, but their absolute standard of living is falling as well. Since the average level of consumption is very low, any fall in income, however small, greatly increases misery.

The rapid demographic increase and the extensive migration of labour from rural to urban areas has thrust an impossible burden on the cities to create employment opportunities and provide essential social services. Yet from Uruguay[4] to Colombia the government has failed to provide minimal services to the agricultural sector, encourage rural areas to retain their labour and productively employ them, or alter land tenure systems. Often the government has done little to assist agriculture other than not impede spontaneous colonization of unclaimed land. This, of course, is not a solution if only because there is hardly any unclaimed land that has immediate agricultural value. In Colombia it was found that even in the so-called national territories large landholdings were prominent. 'An example of the large landholding which exists in the national territories can be found in the eastern plains of Colombia. A landowner who lives in Bogotá owns over 50 thousand hectares in one of the national territories. In addition, he has another large section of land with about 100 thousand hectares in an adjacent national territory.'[5]

Thus the economy of Spanish America is characterized by poverty, resource immobility, slow growth and extreme and worsening income inequality. A large portion of this book has been concerned with examining the impact of foreign economic relations upon the region. We have found that the volume of exports has increased slowly, that Spanish America's share of world trade has declined steadily and

[3] D. Chaplin, 'Industrialization and the Distribution of Wealth in Peru', Wisconsin Land Tenure Center, Research Report No. 18, July 1966, p. 10.

[4] See R. H. Brannon, *The Agricultural Development of Uruguay*, Part III.

[5] Dale W. Adams, 'Landownership Patterns in Colombia', *Inter-American Economic Affairs*, Winter 1964, p. 83.

that throughout the 1950s the terms of trade deteriorated sharply. In the last few years world prices of the region's exports have increased somewhat as a result of the impact of the war in Vietnam and the expansion of demand, particularly for grains, from the communist countries. Nonetheless, the region has borrowed heavily from abroad and the burden of the foreign debt is rising rapidly. These various tendencies have combined to produce an acute shortage of foreign exchange.

Nearly all of the underdeveloped regions of the world have experienced similar difficulties, but the problems in Spanish America in many respects seem to be more intractable than elsewhere. Some Spanish American countries, of course, have fared better than others. These are the fortunate few which happened to export valuable primary commodities or raw materials which enjoyed a rapid growth in world demand. The countries which avoided a foreign exchange bottleneck, e.g. Venezuela in the 1950s and Peru in the early 1960s, were the ones which experienced sustained growth without inflation. The more typical countries, however, are Uruguay (where *per capita* incomes have been falling steadily for a decade), Argentina (where the average rate of inflation recently has been about 25 per cent a year), Colombia (whose terms of trade declined 44 per cent between 1954 and 1962), and Chile (where the ratio of service payments on long-term foreign capital to export earnings was nearly 45 per cent in the period 1961–65).

Spanish America's poor performance is not just due to bad luck; nor as *The Economist* seems to think,[6] is it due simply to a world liquidity shortage and the presence of obstacles erected by the industrial West to the export of manufactured goods from the underdeveloped countries.

It is true that the relative success of those Spanish American countries which have done well can be attributed in part to good luck. It is also true that an increase in world liquidity might induce the wealthy nations to treat exports from the poorer countries much more favourably. Moreover, it is true that the combination of agricultural subsidies in the industrialized nations and the discriminatory tariff structure have hampered Spanish America's exports. But it does not follow that the removal of these barriers would dramatically alter the economic prospects of the region. Spanish America's problems are much more deep-seated than this.

The root of many of the region's problems, it appears, can be found in the past, and particularly in the colonial period. For it was

[6] September 25, 1965, pp. ix–1.

during this time that South America became incorporated into the world economy. It was also during this period that the major political and social characteristics of the region became established. Independence did relatively little to change the economy or the society; its main effect was to transfer power from a mildly benevolent and weak Spanish King to an entrenched creole aristocracy.

This does not imply, of course, that there have been no changes in Spanish America since the colonial period. Quite clearly there have been enormous changes, even in the recent past, e.g. a demographic explosion almost everywhere, intense urbanization, a violent revolution in Bolivia, the development of 'populist' movements in Argentina and Chile, etc. None of the changes which have occurred, however, have been sufficiently strong to alter the nature of the economy or the way in which it responds to stimuli.

The agrarian structure evolved gradually from the beginning of the Conquest and was not consolidated in its present form until the middle of the last century. The latifundia system which we see today is, of course, still changing, but it is doing so at a rate which is too slow to meet the demands placed upon it. Likewise, the structure of international trade changed from time to time in response to changing conditions abroad and the discovery of new raw materials within the region. The present pattern of trade, and in particular the close links with the United States, is largely a product of the first third of the present century.

In the course of time the separate Spanish American nations have become highly specialized in the export of one or two primary commodities. This process is both reflected in and caused by the extension of the latifundia system, the marked inequalities of income, and the exploitation of one group by another. The high degree of specialization, in turn, has led to a rigid structure of production and the fragmentation of the economy. The independent nations of Spanish America became vulnerable to business cycles originating abroad and to the pronounced technical dynamism of the industrialized countries. This vulnerability is reflected in part in the oscillations of their terms of trade and in the slow growth of their exports. The result of Spanish America's integration into the world economy and the present international division of labour has been to hinder the emergence of a vigorous indigenous industrial entrepreneurial class, to weaken the ability of the economy to transform resources and to reduce the capacity of the society to accept change.

The trading arrangements and policies of the industrialized nations simply exacerbate the existing structural deformity. Marginal changes in policy (which is the most one could expect) would not be sufficient radically to improve the development prospects of the region. They would certainly be better, however, than the present arrangements in which the pernicious effects of the trading policies of the rich nations are 'compensated' by capital flows which—even when they are on balance favourable to the poor nations—are conditional upon shortsighted political agreements, the payment of interest, and the freedom of foreign businessmen to monopolize important sectors of the economy and reap extraordinary profits.

The implications of this analysis are, perhaps, paradoxical. The past performance of the Spanish American economies is a direct reflection of the way in which this region was incorporated into the world economic system by the Spanish Conquest. The relationship between Spanish America and what is today the industrialized West was unequal from the very beginning and has remained so. In this sense 'foreign trade' and the 'international division of labour' bear some of the responsibility for generating and sustaining a state of economic underdevelopment and social backwardness in the region. It would be wrong to suppose, however, that a marginal change in the trading arrangements between the individual Spanish American countries and the rest of the world would by itself be sufficient to rectify the consequences of an inherently unequal association. Spanish America's problems may be international in origin, but the solution of these problems must be largely national and regional.

The reason for this is that it is not the ill will of men but the inadequacy of the economic system which perpetuates poverty in the area. For example, a latifundia system, by its very nature, constitutes a low-productivity economy and an unequal income-distribution-society. The difficulties of such an economy cannot be solved, say, by obtaining a higher price for wheat, but only by altering the system itself. It is for this reason that throughout this study stress has been placed on structural rather than marginal changes. The essence of development in Spanish America consists in converting the constants of marginal analysis into variables.

Historically this seems to have occurred most frequently when the ties of the region to the world economy were severed. Social, political and economic constants often have been converted to variables when the economy of the region has been subjected to apparently severe blows caused by major international wars and depressions. Professor Kafka has noted the fact 'that relatively rapid progress not only of structural change, particularly of industrialization, but also of overall growth have sometimes been associated with violent adverse shocks to the economic system'.[7] Professor Hirschman has made a similar observation: 'Wars and depressions have historically no doubt been most important in bringing industries to countries of the "periphery" which up to then had firmly remained in the nonindustrial category.'[8]

In these periods of crisis in the world economy Spanish America flourished; and the deeper the crisis the more the region bloomed. 'In the Great Depression, there appears . . . to have been a curious association between the degree of violence of these adverse shocks and growth.'[9] It was when contact between the industrial West and the region was broken that Spanish America grew rapidly. It was when the region was deprived of imports, foreign technical expertise and external capital that it was able to become more industrialized. It was then that urbanization proceeded swiftly, that the State assumed new responsibilities and functions, and it was then that social measures in housing, education and health were first introduced.

The severing of customary ties between the centre and the periphery, the breaking of commercial contact between the industrial and the nonindustrial nations, and the weakening of external political influence in periods of world crisis enabled Spanish America to begin to reverse the process of underdevelopment—often only to fall back again once the crisis was over. In some respects a world war or depression is to Spanish America as a tariff is to a nascent industry: it interposes a barrier between the strong and the weak, enabling the latter to grow and develop, to acquire experience and strength. Just as an industry will wither if the tariff is removed too soon, so a poor country will fall back into underdevelopment if the crisis ends prematurely. It is not the 'shock' of a world crisis that stimulates development in Spanish America, but the protection it affords to infant industries, the incentives it provides to alter policies and the freedom it gives from foreign interference.

To revert to one of the themes of the Introduction, foreign contact between two grossly unequal economies may help to create underdevelopment, while the lack of contact may permit—but does not ensure—development. How is it possible that peaceful contact

[7] A. Kafka, 'The Theoretical Interpretation of Latin American Economic Development', in H. S. Ellis, ed., *Economic Development for Latin America*, p. 8.

[8] A. O. Hirschman, 'The Political Economy of Import-Substituting Industrialization in Latin America', *Quarterly Journal of Economics*, February 1968, p. 4. Also see A. G. Frank, *Capitalism and Underdevelopment in Latin America.*

[9] A. Kafka, *op. cit.*, p. 9.

between the rich and the poor works to the disadvantage of the latter? Part of the answer to this question is scattered about the earlier chapters of this book. Let us recapitulate our findings.

First, contact between the rich and the poor may foster discontent and undermine social cohesion. The 'demonstration effect' may stimulate aspirations for consumption and create a desired level and pattern of demand that cannot be satisfied by the economy's ability to produce. Furthermore, the importation by labour abundant economies of techniques of production designed for capital abundant economies may lead to lower output, slower growth, greater unemployment and more inequality than would be necessary with more labour intensive techniques. Secondly, contact may lead to the adoption of institutions that are inappropriate to the conditions of the poor countries. In some instances these institutions may have been imposed on the poor, e.g. the latifundia system. In other cases, these institutions may have been imitated by the governments of the poor, e.g. the introduction of public consumption programmes for the middle class, appropriate to a welfare state, in countries where the volume of production is insufficient to justify them or permit their extension to the majority of the population.

Third, contact and competition between unequal economies leads inevitably to a pattern of trade which is neither free nor organized in the interests of the periphery nations. Subsidies in the rich nations on temperate climate agricultural products, protection of domestic sugarbeet growers, import quotas on textiles and petroleum, the tariff structure on manufactured goods are all examples of devices which hinder the industrialization of the poor countries and restrict the markets for some of their primary commodity exports. As a consequence, the underdeveloped countries become highly specialized in the production of a few products (for which long-run prospects usually are not good) and their economies become inflexible and incapable of responding rapidly to changed conditions. The resulting division of labour is one in which the poor countries tend to specialize in extractive industries, tropical agricultural products and technologically stagnant manufactures, while the rich tend to concentrate on those industrial activities which generate widespread external economies and those which are characterized by rapid technical progress.

Just as the distribution of the benefits from trade in commodities may be asymmetrical—favouring the industrial nations more than the underdeveloped countries—so the growth and exchange of knowledge and information may favour one type of country more than another. This is the fourth way in which contact between the rich and the poor may cause the underdevelopment of the latter. The most obvious example in Spanish America is the importation of medical knowledge from the West which led to a sharp fall in the death rates and a resulting population explosion. At the same time, the previous importation of an institution—Roman Catholicism—hinders the application of other knowledge, also available from the West, which would enable the region to regulate births.

Finally, there is the tendency of dynamic countries (and regions within countries) to attract factors of production away from the underdeveloped countries and regions. The rich, technologically progressive, capital abundant, rapidly growing economies exert a strong pull on the capital of the underdeveloped countries. Contrary to neo-classical economic theory, capital often flows from the capital-scarce to the capital-abundant countries, and not the other way around. Capital movements, rather than being equilibrating, are disequilibrating, setting up cumulative movements which widen income disparities between the rich nations and the poor. Indeed, there is considerable evidence that on balance Spanish America transfers capital to the industrial West—despite foreign aid and government controls to prevent the export of private capital.

As we argued in Chapter III (and by implication in Chapter IV), additional capital imports is not the answer to this problem. In fact, more foreign aid and direct, private foreign investment in Spanish America could easily make matters worse rather than better. This is because capital imports tend to lower the domestic savings rate, foreign private investment tends to retard the emergence of an indigenous entrepreneurial class, foreign aid often distorts the pattern of investment and leads to a higher capital-output ratio, the servicing of foreign capital ultimately creates severe balance of payments problems and because foreign capital tends to inhibit structural reform and perpetuate the *status quo*. One cannot reverse the present tendency for capital to flow from Spanish America to the industrial West by pleading for more aid (which in any case would be futile) or by offering more attractive investment codes to foreign capitalists (who already are given more favourable treatment than local investors). A reversal of the flow will require major changes in the behaviour of the system.

Capital, moreover, is not the only factor of production which is attracted to the industrial nations. Skilled labour, but not unskilled labour, is highly mobile, moving from the poor countries (and regions within countries) to the rich, in response to the higher salaries the latter can offer. The international migration of human capital—the so-called brain drain—has been a subject of discussion and debate for quite some time. Orthodox economists have concluded that the migrants and the country of immigration gain from this movement, while the country of emigration, in general, does not lose, and may even benefit, e.g. from emigrants' remittances. Thus the orthodox economist argues that the brain drain increases efficiency, welfare and output.[10]

The conclusions, however, are valid only if one assumes that changes are marginal, that labour is paid the value of its marginal product, that there are no external economies and thus private costs accurately reflect social values and, finally, that the distribution of income does not matter. One has only to state the assumptions to realize how inapplicable they are in Spanish America and other underdeveloped countries.

First, salaries do not reflect marginal product but are determined to a large degree by the presence or absence of monopoly power, restrictions to entry into the professions, conventional salary structures and the like. Secondly, highly skilled labour, technicians and scientists generate external economies; they are the people largely responsible for introducing and spreading new ideas, changing attitudes and reforming institutions. Thus the monetary value of their salaries in no way reflects their contribution to the social product.

Third, the emigration of human capital from Spanish America is not a marginal phenomenon which leaves unchanged the *per capita* income of the unskilled labour left behind. On the contrary, the brain drain tends to lower the income of those who remain. One reason for this is that highly trained and able people are terribly scarce (and hence rather 'lumpy'). The emigration of a single, key individual may deprive his colleagues 'of stimulus, effectiveness and even of employment opportunities'[11]. The loss of the director of a demographic research institute may lead to the closure of the institute; the emigration of a senior chemist may greatly affect the efficiency of an important enterprise; the departure of the chief economic planner may disrupt the planning agency and reduce the effectiveness of the government's development policy, etc.

Another reason why emigration lowers the average income of

[10] Harry Johnson's contribution to the symposium on *The Brain Drain* (edited by Walter Adams) is a good statement of the orthodox position.

[11] Paul Streeten, 'Training and Draining Brains', *New Society*, July 11, 1968, p. 59.

those who remain is that it is occurring inside (and not on) the margin.[12] Assuming diminishing marginal product of labour and a wage equal to the marginal product, the departure of one person will lower national product only by the amount of the emigrant's wage, leaving the average income of the rest of the population unchanged. However, the emigration of any additional workers will reduce the average income available for the remaining population, because the departure of 'the second and each succeeding emigrant will reduce national product by an amount which is greater than the income which they had been receiving in the initial situation'.[13]

This can be understood quite easily with the help of simple notation. Assume α is equal to the marginal product of the 'n'th worker, which is equal to the wage. Assume also that the contribution of the $(n - 1)$ worker to national product $= \alpha + k$, and of the $(n - 2)$ worker $= (\alpha + k + k)$, etc. If one worker emigrates national income and output fall by α, the amount of the wage, and the average product available for those left behind remains unchanged. If two workers emigrate, national income falls by $\alpha + (\alpha + k)$. The claim of the two hypothetical emigrants on national product, however, was only their wages $= 2\alpha$, leaving a surplus, k, to be distributed among the rest of the population. This surplus vanishes upon emigration and in consequence the rest of the population is made worse off by this amount. Similarly, if three workers emigrate the income available for the remaining population declines by $3k$ $[= \alpha + (\alpha + k) + (\alpha + k + k) - 3\alpha]$. Thus the brain drain does lower the real income of those left behind.

Furthermore, the brain drain leads to a redistribution of income within the country of emigration in favour of technical and skilled workers. As these workers become even more scarce their marginal product, and hence their wage, will rise. Their incomes, in other words, increase absolutely, despite the fact that the average income of all those left behind declines. This redistributive process further lowers the income of the unskilled workers. That is, emigration of skilled workers harms the unskilled workers in two ways: by lowering the average income of those who remain and by redistributing

[12] One (inadequate) indication of this is the fact that 11·9 per cent of all economists in the United States were born abroad. See H. G. Grubel and A. D. Scott, 'Foreigners in the US Economics Profession', *American Economic Review*, March 1967. Some data for Latin America are available in G. R. Gonzales, 'The Migration of Latin American High-level Manpower', *International Labour Review*, December 1968.

[13] N. D. Aitken, 'The International Flow of Human Capital: Comment', *American Economic Review*, June 1968, p. 540.

this smaller income in favour of skilled, technical and professional workers.

Thus it is that mere contact between rich and poor countries works to the disadvantage of the latter. Through the imitation of patterns of consumption and techniques of production, through the importation of inappropriate institutions, and through the normal flow of commodities, knowledge and factors of production the outcome of (largely) unrestricted economic intercourse between Spanish America and the industrial West is to accentuate and perpetuate inequality and underdevelopment. This does not mean that the conscious exploitation characteristic of a colonial relationship has not occurred; it means only that one need not explicitly introduce it in order to explain the underdevelopment observable in Spanish America. Once a nation or region becomes underdeveloped, for whatever reason, strong economic forces tend to keep it there. It is contact between unequal partners of the world community that prolongs underdevelopment, and it is in periods when this contact is broken that the first steps toward development often begin.

Of course there is no suggestion that those interested in the economic development of Spanish America ought to advocate world wars and major international depressions. Such events are unpredictable, arbitrary and wasteful; they cause unnecessary human suffering and, in any case, cannot be controlled from within the region. But if our analysis is correct development in Spanish America would be accelerated if regulation of international economic intercourse were coupled with purposeful national planning, social reform and political reorientation.

Nationally, the most urgent economic requirements are a complete overhaul of the tax system and a sweeping agrarian reform. In every country collection procedures, tax rates and the tax structure will have to be altered. It is essential to raise the rate of investment, improve its allocation and increase expenditure on technical education and applied research. The only way to do this is through generally higher taxes and a more efficient and equitable tax system. At present, government net savings are an extremely small proportion of the national income. For example, they are about five per cent in Ecuador, three per cent in Chile, two per cent in Colombia and zero in Bolivia; only in Venezuela, where government savings are about nine per cent of national income, is the saving effort of the government not derisory.[14] Throughout Spanish America the rate of capital accumulation must be sharply increased.

[14] UN, *World Economic Survey 1965*, 1966.

An agrarian reform is necessary in order (a) to increase food output and raise the nutritional level of an undernourished population, (b) to provide a more equitable distribution of income and wealth, and (c) to increase the size of the domestic market for mass produced manufactured goods. Changes in land tenure arrangements would not by themselves be enough to achieve these objectives. A land reform would have to be reinforced and complemented by a series of other measures designed to ensure that the other inputs (including knowledge) which modern agricultural production requires were available at the right moment, in the appropriate qualities and in sufficient quantity. Furthermore, steps will have to be taken to ensure that the resulting output could be efficiently stored, transported and marketed. In other words, the over-all management of the agricultural sector will have to be altered.

The major thrust of development, however, must occur in the industrial sector. The possibilities of increasing the production of primary commodities for export are limited by the relatively slow rate of growth of world demand. Large increases in the output of primary commodities would in most cases simply result in lower export prices and an even greater specialization in products for which the long-run prospects are poor. Unfortunately, industrialization through import-substitution at the national level has been inefficient, due in part to the small size of domestic markets and the consequent inability of producers to take full advantage of economies of scale. Further possibilities in this direction have nearly been exhausted. At the same time, production of manufactured goods for export to the industrial nations is extremely difficult, but not impossible. This is because the Spanish American countries are high cost producers and are likely to remain so until they have acquired experience and knowledge in industrial techniquès, trained additional skilled labour, and established profitable commercial relations with potential buyers.

As indicated earlier, most of the industry in Spanish America arose when contact with the industrial West was temporarily diminished. The growth of manufactures 'started with relatively small plants administering "last touches" to a host of imported inputs, concentrated on consumer rather than producer goods, and often was specifically designed to improve the levels of consumption of populations who were suddenly cut off, as a result of war or balance-of-payments crises, from imported consumer goods to which they had become accustomed'.[15] Import controls and tariffs—now

[15] A. O. Hirschman, *op. cit.*, pp. 8–9.

the highest in the world—originally were used to keep out 'non-essential', i.e. luxury, consumer goods. Since the tariffs were not matched by internal taxes on locally produced substitutes, the trade controls provided strong incentives to establish local industries producing luxury consumer items. In consequence, industrialization became an irrational process which reflected neither a country's costs of production nor its needs for 'essential' consumption goods.

Moreover, looking back over the last few decades, industrialization in Spanish America appears to have been a triple failure: it has not solved the region's problem of unemployment; it has not economized on foreign exchange and eased the balance of payments problem;[16] and, finally, it has not succeeded in transforming the economy and society of the region. About the most that can be said in its favour is that it has not been difficult. Industrialization in Spanish America, perhaps surprisingly, has left the region substantially unchanged.

Socially, those who organized and ran the new manufacturing establishments usually came from the upper classes or joined them from abroad. For instance, in Peru, 'Of the 45 family and corporate entities on the Board of Directors of that country's *Sociedad Nacional de Agricultura*, 56 per cent are important stockholders in banks and financial companies, 53 per cent own stock in insurance companies, 75 per cent are owners of companies engaged in urban construction or real estate, 56 per cent have investments in commercial firms, and 64 per cent are important stockholders in one or more petroleum companies'.[17] Thus the indigenous Peruvian industrial and latifundista classes overlap to a considerable degree. As regards the foreigners, many of them arrived as representatives of large foreign firms which were anxious to maintain their market in Peru and therefore were willing to shift from exporting to import-substituting manufacturing. Others were immigrants starting business on their own. In fact, the great majority of entrepreneurs in Peru's industrial sector have been foreign immigrants. It is noteworthy that 'in general, white immigration to Peru has been small, probably not over 20,000 since 1900. However, these foreigners, Anglo-Saxon and Italian immigrants alike, arrived—not as marginal strangers—but elite entrepreneurs—moving in near the top and usually consolidating their positions, thus making Peru, while nominally politically independent, very much a cultural and economic colony'.[18]

[16] In this regard see the interesting article by Carlos Diaz-Alejandro, 'On the Import Intensity of Import Substitution', *Kyklos*, 1965.

[17] A. G. Frank, *op. cit.*, p. 113.

[18] D. Chaplin, *op. cit.*, p. 24.

The question now arises as to whether the process of industrialization experienced by Spanish America should be pushed further, perhaps placing more emphasis on exports of manufactures to the world market and less on import substitution for the domestic market. Professor Hirschman believes that this process can and should be pushed much further.[19] On the other hand, if industrialization in the past has failed to become a spearhead for development and a catalytic agent transforming Spanish America, is there any reason to believe that continuation of the same strategy will be successful in the future? Indeed, if the separate countries continue to pay no heed to long-run comparative costs and if they enter into a large number of new activities in which the technology is vastly more complex, the initial investment requirements are greater and the importance of economies of scale is increased, a continuation of past policies may lead to considerable disappointment.

What seems to be needed is a policy for industrialization that concentrates on regional planning and the import substitution of capital goods industries. Such a strategy, of course, presupposes and is contingent upon a high domestic savings effort and the ability of the government to mobilize an agricultural surplus. Thus the industrial strategy must form part of a much wider programme for development. It is in this context that a regional scheme for industrialization, as presented in the last chapter, becomes an attractive alternative. It would ease the foreign exchange bottleneck facing most of the Spanish American countries and it would allow each of the participating nations to establish at least one industrial complex. Furthermore, the possibility of efficient production would be encouraged by the formation of a large regional market in which economies of scale could be fully exploited.

The size of the regional market, of course, depends on many factors in addition to the geographic area encompassed: the size of the population, population density and the adequacy of transport services, *per capita* income and income distribution. Taking these factors into account, as well as export possibilities, it is clear that small domestic markets hamper the development of many producer goods industries in Spanish America. As Professor Chenery has remarked,'. . . at the income level of $300 . . . economies of scale are probably significant up to a population of 100 million or more'.[20] Regional planning of industry could help Spanish America exploit

[19] *Op. cit.*

[20] H. B. Chenery, 'Patterns of Industrial Growth', *American Economic Review*, September 1960, p. 646.

economies of scale in a large number of producer goods, e.g. cement; oil refining; production of oxygen, synthetic ammonia and other chemicals; aluminium, iron and steel making plus other metals; transport equipment, rubber and paper. The existence of a large regional market and the installation within it of adequate capacity in the capital goods industries would enable Spanish America's growth rate, for the first time, to become independent of the growth of world demand for primary products.

The current attempts at regional integration within LAFTA are completely inadequate. In fact, these efforts have barely been sufficient to re-establish the relative importance which intra-Latin American trade enjoyed during the Second World War and the period of bilateral agreements in the mid-1950s. Indeed, it was only in 1964 that the value of intra-LAFTA trade surpassed that of 1955. As presently conceived the Latin American Free Trade Association is likely to benefit hardly anyone other than large foreign enterprises and the biggest member countries—Brazil, Mexico and, possibly, Argentina.

To those who believed that the Alliance for Progress constituted a radical initiative in international economic relations, the April 1967 'Declaration of American Presidents'—following the meeting at Punta del Este, Uruguay—must have come as a shock. For it was at this meeting that the Heads of State of the countries participating in the Alliance gave top priority to the formation of a common market, and in effect repudiated the basic premise of the Alliance that integration would follow and not precede fundamental reforms in agriculture, education, politics and taxation. Tariff reductions have now been put forward as a substitute for structural change, and regional integration has become another prop to the *status quo* rather than an instrument for accelerating capital formation and industrialization in a more egalitarian society.

The development strategy advocated in this chapter would, of course, require a high degree of national planning and regional co-ordination. Without it, policies are likely to be contradictory and inadequate; savings will be insufficient to finance capital investment; the new units of production may be unable to hire enough skilled labour; industrial plants may be duplicated in several integrating nations; markets may not be found for the additional output, etc. The measures recommended cannot be implemented merely by changing relative prices, paying subsidies or lowering tariffs; implementation will also require a series of institutional reforms, direct controls and administrative decisions over the allocation of resources.

To be efficient, implementation must be systematic and rational, i.e. planned.[21]

At present none of the Spanish American countries are in a position to implement the recommended measures, because their administrative machinery is quite inadequate. A recent study published by the Economic Commission for Latin America indicates that there is no planning at all in Paraguay and Uruguay; only partial investment programmes in Peru and Argentina; and long-term plans without any programme for financial implementation in Bolivia, Chile, Colombia, Ecuador and Venezuela.[22] This in itself, although disappointing, is not very surprising—because a decision to plan in Spanish America would be inextricably linked with a decision to plan *for something*. Whatever the 'for something' might be, it is unlikely that a complex planning apparatus would be created merely to perpetuate the *status quo*, although this is not inconceivable, of course.

At present the activity that is called planning consists in large part of the preparation of macro-economic projections used as supporting documents for requests for foreign assistance. Thus planning is largely an exercise in diplomacy—undertaken at the behest of AID, with technical assistance, in order to get aid. Genuine planning, in contrast, has internal political implications; it would require substantial adjustments in the political organization and in the distribution of power in the society. For example, effective planning would almost certainly require some representation of working class interests. Thus the introduction of systematic planning in the region would not only be an exciting event in itself; more importantly, it would be an indication of a fairly substantial change in ideas. In particular, planning, nationalism and demands for popular participation in the decision making process are likely to go together. They would constitute an affirmation of national identity, an assertion of national sovereignty and an avowal of democracy and equality as national objectives. Thus economic planning and political nationalism would be as much a symptom of structural change as a cause.

Throughout this study of underdevelopment in Spanish America we have been forced to refer to political issues. In the final analysis the region's poverty is a reflection of the politics of the past. In a sense, an attempt now to develop in Spanish America depends upon political processes (and whether they will be violent or evolutionary

[21] For a contrary view see A. O. Hirschman, *Journeys Towards Progress*.

[22] UN, ECLA, 'Progress in Planning in Latin America', *Economic Bulletin for Latin America*, October 1963.

is unknown) as much as upon economic decisions. Moreover, whether an attempt to develop succeeds depends as much upon international political attitudes as it does upon market conditions in the world economy. The profound changes necessary for social and economic progress in Spanish America will inevitably impinge upon the interests of powerful groups in the industrial West as well as upon the privileged classes in the region itself. Development will require not only internal reforms; it will also require alterations in the international economic system, for it is in part the working of this system which prolongs underdevelopment in the region.

Not all changes are opposed, of course. Indeed, the ruling groups and their Western supporters may view some change as a necessary concession to popular demands and as a legitimate exercise of national sovereignty. The more fundamental changes, however, usually encounter formidable resistance and are attacked as an illegitimate threat to the economic and political wellbeing of privileged local and foreign groups. Economics often is used as a weapon by those who oppose radical reform and economic arguments regularly are adduced to support the *status quo*. If nothing else, I hope to have shown in this book that many of the economic arguments used by those who dislike change are specious. This does not imply that reforms can be implemented without cost. Indeed, the economic costs of reforms sometimes are high, as we are frequently reminded by those who have much to lose; but it must also be recalled that the benefits of reform can be considerable, particularly for those who have most to gain.

Index

123794

LINKED